Air Sampling
and
Industrial Hygiene Engineering

Air Sampling
and
Industrial Hygiene
Engineering

Martha J. Boss
and
Dennis W. Day

CRC Press
Taylor & Francis Group
Boca Raton London New York

CRC Press is an imprint of the
Taylor & Francis Group, an **informa** business

CRC Press
Taylor & Francis Group
6000 Broken Sound Parkway NW, Suite 300
Boca Raton, FL 33487-2742

First issued in paperback 2019

ISBN-13: 978-1-56670-417-5 (hbk)
ISBN-13: 978-0-367-39771-5 (pbk)

Library of Congress Cataloging-in-Publication Data

Boss, Martha J.
  Air sampling and industrial hygiene engineering / Martha J. Boss, Dennis W. Day.
    p. cm.
  Includes bibliographical references and index.
  ISBN 1-56670-417-0
    1. Air—Pollution—Measurement. 2. Industrial hygiene. 3. Air sampling apparatus. I. Day, Dennis W. II. Title.

TD890 .B66 2000
628.5′3′0287—dc21
                                                                                    00-048666
                                                                                    CIP

Library of Congress Card Number 00-048666

Visit the Taylor & Francis Web site at
http://www.taylorandfrancis.com

and the CRC Press Web site at
http://www.crcpress.com

Preface

Many have endeavored to make our outdoor environment cleaner and safer. The learning process that occurred showed us the limitations of our planet and also the sustainability of our ecosystem if given a chance. As a community, we learned about the water, the soil, and the air. We learned about the underground river that flowed to the surface lake. We learned about air currents that transported airstreams around our globe. We discovered the reality of plate tectonics and the ever-changing hydrogeological system. Using this knowledge, we continued to learn how to clean our environment and prevent further damage.

Our science careers began with teaching and working on environmental issues. During that time our concern for 1 ppm benzene at an underground storage tank (UST) location was intense. Then as we learned more, we began to see what had been invisible to us before—the air in our factories, hospitals, schools, homes, and cars. We began to realize that environmental concerns and our accumulated knowledge on how to protect people and the environment was not being translated into knowledge about buildings in which people live and work. Many people routinely work in factories where exposure to hundreds of parts per million of benzene is commonplace.

Six years ago we received a call from a farm family in the Midwest. For three generations they had farmed their land. Now their children, their farm animals, and they themselves were sick. A chemical storage fire had burned out of control and covered their land and homes with oily soot. Yet that spring they planted their fields and tried to live their lives as before.

As the planting season progressed, farmers sickened in the fields. Upon returning to their homes, the sickness increased. The vehicles they used in the field became more and more contaminated. The farmers began buying old cars and abandoning them when they could ride in them no longer. Two combines were also abandoned. They left their homes, in some cases the original farm homesteads that had housed three generations.

Planting was over and the hogs were farrowing. The animals were born deformed; the mother animals died. Eventually most of the animals sickened and were sacrificed. The farmers began looking for answers.

Fall approached and with that the harvest. The farmers reentered the fields and became increasingly sick. What to do? Should they even harvest these crops? Should their children be sent away?

Winter came—was it all in their imagination? The doctors and scientists they had contacted were without answers. Perhaps it would be better in the spring.

Spring arrived, planting began, and the cycle continued. From somewhere, they were given our name. We arrived and began investigating. These farmers and their families had not benefited at that time from the collective knowledge available pertaining to fires and chemical dispersion. Particulates laced with chemicals can exit the periphery of a firestorm. The chemicals could remain intact or even recombine. Many chemicals can remain in our soil and water; after plowing with combines, these chemicals reenter the airstream and become available once again for us to breathe and carry home on our clothing.

These farmers were carrying home the vestiges of chemicals we use as pesticides and herbicides. Chemicals that had changed in the fire became more toxic at lower levels. Chemicals were rendered more easily available by their current adsorption to airborne soil particulates. Upon entry of these particulates into their lungs, the new chemical mix off-gassed and became biologically active. In the heartland of America, these farmers had unwittingly participated in an experiment in chemical warfare!

We decided then to write a book to open a dialogue on air monitoring, risk, and engineering—a book to show that collectively we as scientists and engineers need to develop an interdisciplinary approach to applying our knowledge.

Before any art must come the science. Chapter 1 (Air Sampling Introduction), 2 (Air Sampling Instrumentation Options), and 3 (Calibration Techniques) present the current state-of-the-art techniques for air sampling. Chapter 4 discusses statistical analysis and relevance issues.

In Chapters 5 (Chemical Risk Assessment) and 6 (Biological Risk Assessment), we discuss how air sampling and other environmental sampling are used to determine risk—risks of acute effect, chronic effect, and carcinogenic effect. Biological risk always has the added element of reproduction, as biologicals, unlike chemicals, can enlarge their numbers over time and distance from their source.

We then turn our attention to Chapter 7 (Indoor Air Quality and Environments) and Chapter 8 (Area Monitoring and Contingency Planning). Once we know how to monitor potential risk, how do we evaluate our buildings, our city air, and all the places we live and work? What do we do in an emergency? Are there times as illustrated in Chapter 9 when we will need to use microcircuitry and remote monitoring? What about our workplaces as addressed in Chapter 10 (Occupational Health—Air Monitoring Strategies)?

Finally we need to consider monitoring for toxicological risk (Chapter 11). If we find risk is evident, what tools (Chapter 12, Risk Communication and Environmental Monitoring) will be needed?

This book is the start of an interdisciplinary look at many issues that in fact are just one—can we live and work in places that are healthy? Do we have the knowledge and resources to ensure that our hospitals and schools have clean air? Can we now build and maintain ventilation systems that do not foul over time?

After World War I, Martha's grandfather returned to work in a cement plant. He was having some trouble breathing after he inhaled mustard agent in the trenches of France. At the cement plant he dug into the earth at a quarry using shovels and eventually powered equipment. The dust swirled around him and coated his clothing. Every night he was racked with convulsive coughing. In the morning he felt better, could even smoke on the way to work. Over the next 30 years, he slowly died. No one knew then to tell him—get another job, quit smoking, protect your damaged lungs.

Dennis's father was a plumber. He watched pipe fitters carry buckets of gray slurry to the work site. The slurry was applied to pipe junctures and hardened to ensure pipe integrity. The pipe fitters used their hands and wiped the excess slurry on their clothing. They returned home, where their clothes were washed with their family's clothes; often the laundry room was next to the air intake for their home furnace. Over the years Dennis's father watched all these men die as their lungs, scarred with asbestosis, failed.

How many men and women to this day still do not know that the factories and workplaces they occupy are poisoning them and often their families? Do they not know because the knowledge is unavailable? No. However, we have been slow to realize the need to communicate our knowledge. The simplest concepts have been lost. You do not have to die to work. Ventilation systems can be improved. Healthier workers are more productive workers and happier people.

As our buildings age, and as we use ventilation systems designed to heat buildings—and to cool them—our indoor air problems have multiplied. The heat and cool cycles often cause condensation within the air-handling systems. The fiberglass duct liners that have captured particulates become slightly wetted. With time molds and fungi begin their life cycles hidden from us and amplify in number. Their spores ride the duct's airstream to our rooms and hallways. Maintenance personnel cannot reach the biological hiding ground.

Our residents begin to notice their health decline. Biological risk? Yes. In our hospitals and schools? Yes.

Our hope is that this book will be used to begin these dialogues. Engineers and scientists need to look holistically at building design and maintenance. Business people need to realize the financial risk associated with accepting a nice building front rather than a state-of-the-art ventilation system. We all need to begin talking and learning together, so that our children can live and work without concern for the very air that they breathe.

Martha J. Boss
Dennis W. Day

About the Authors

Martha Boss is a practicing industrial hygienist and safety engineer living in Omaha, NE and various airports throughout the United States. Many years ago, Martha won the Army Science award at the Des Moines, IA science fair. As fate would have it, Martha eventually worked for the Army and through the auspices of EPA grants was trained in industrial hygiene. All of this surprised Martha because she had intended to teach high school science and had prepared herself for that endeavor with a B.A. in biological education (University of Northern Iowa) and later a BS in biology (University of Nebraska).

During Desert Shield that became Desert Storm, Martha was tasked under the War Powers Act to assist in the preparation of a western Army base to house and train special forces. Dennis was also so commissioned, and their professional association began.

Martha worked with her fellow Army industrial hygienists and engineers to assess biological, radiological, and chemical warfare sites and find solutions. The Army continued her training at such institutions as Johns Hopkins, Harvard, and other top centers throughout the nation.

After five years of traveling throughout the country to various very scary places, Martha decided to settle down in a regional engineering firm. After a couple of years, Martha realized she did not want to settle down and joined a national engineering firm where she is employed to this day. Martha is a principal toxicologist for URS Corporation and continues her practice as a certified industrial hygienist and certified safety professional (safety engineer). Martha is a member of the Hazardous Substances Research Center T^3 board for Region 7 of the EPA, a diplomate of the American Academy of Industrial Hygiene, serves on the editorial advisory board for Stevens Publishing, and is a member of the American Industrial Hygiene Association and the American Society of Safety Engineers.

Dennis Day is a practicing industrial hygienist and safety engineer living in Omaha, NE and various airports throughout the United States. Dennis began his career as a forester. For several years, he traveled through the forests of the East and South cruising timber. Then he decided to become a high school science teacher. Dennis used his B.S. in forestry (University of Missouri) to enable him to pursue additional studies in chemistry and biology (Creighton University) and become a professional teacher. After teaching for awhile Dennis was persuaded to join the Army Safety Office and ultimately the Omaha District engineering division.

Dennis continued for ten years to work with various Army, EPA, and Department of Defense missions. His work included sites throughout the nation and in Europe. Dennis concentrated his efforts on streamlining site assessment protocols, community outreach with protective action plans for chemical warfare sites, and training industrial hygienists entering the Army work force.

Eventually, Forrest Terrell of Dames & Moore (now URS) convinced Dennis to join that firm to develop an interdisciplinary industrial hygiene, safety, and engineering service to commercial and governmental clients. Dennis is a principal toxicologist for URS Corporation and continues his practice as a certified industrial hygienist and certified safety professional (safety engineer). Dennis is a diplomate of the American Academy of Industrial Hygiene and a member of the American Conference of Governmental Industrial Hygienists, the American Industrial Hygiene Association, and the American Society of Safety Engineers. In 1992 Dennis received the Achievement Medal for Civilian Service for his emergency industrial hygiene support following Hurricane Andrew.

Contents

CHAPTER 1

Air Sampling Introduction

This chapter provides an overview of air sampling and site monitoring that is legally defensible. It answers questions about monitoring protocols that must be initiated for emergency and contingency situations.

1.1 DOCUMENTATION

Essentially in any sampling endeavor, without documentation, you have what is called "personal opinion." The intent of documentation is to provide the basis for professional opinion. The documentation then becomes a dialogue of historical perspective and the empirical sampling event.

When assembling historical documentation, make sure that you define how the information is used. Often in order to understand sampling results, the environment, including work practices, must be analyzed. Work practices include human factors; therefore, you must be very careful to present a dialogue of these work practices that is not individually invasive.

When you decide on documentation techniques, before you write the first chronicle, enter field notes, or take that first picture, make sure that everyone understands what the purpose of the information is and who will have access to this information.

Be very careful when the original scope of work calls for only general information. Often as the investigation continues, you may be asked to provide very specific reasons for general information development. Consequently, even for general information screening or limited audits, you may need to keep very specific information available to substantiate your opinions.

Whenever a review of human factors is required, especially interviews, all parties must understand the limitations on personal anecdotal information. If you do not intend to personally name the interviewee, make sure all parties understand.

The overriding message here is define your scope of work and write this definition for all to see. If you need to work under the auspice of attorney/client privilege or through any other set boundaries, make sure the scope of work reflects these facts.

1.2 SAMPLE DOCUMENTATION

To assist in determining appropriate engineering controls, take photographs (as appropriate) and detailed notes concerning the following:

- Visible airborne contaminants
- Work practices
- Potential interferences
- Movements
- Other conditions

Prepare blanks during the sample period for each type of sample collected. One blank will suffice for up to 20 samples for any given analysis. These blanks may include opened, but unused, charcoal tubes.

1.3 COMPETENCY FOR SAMPLING TECHNICIANS

When deciding who is defined as a competent sampling technician, the first criterion is as follows: Is the scope of work completely defined according to sampling requirements? If you have any doubts about the situation with which samplers are involved, whether those doubts stem from a lack of background knowledge of the site or unknown hazards, the sampling scope is not completely defined.

For undefined sampling scopes, a senior sampling professional will need to initiate the site work. If the sampling choices once in the field are multifactorial, in that circumstances on-site are very dynamic, a team of senior sampling professionals is required.

Remember that a phenomenon known as perceptual shift will occur during sampling. As we become more or less secure with our environment, we start to see things differently. A strong team is able to keep its members on target, thus providing a more complete picture of the sampling environment.

Once a scope of work is defined, sampling can often be delegated to less senior personnel. The purpose of this text is to provide information to assist in the standardization of both the initial and the delegated work effort. Despite the many ways of communicating before the sampling event, dialogue must be continued throughout the sampling event. Unfortunately this dialogue may not be free-flowing conversation without limitations. The original team that defined the scope of work must be available to the on-site personnel. The actual conversations on-site, while delimited by many events, must give way to free-flowing discussions within this team so that the collected data are useful and relevant.

1.4 SAMPLING ACTIVITY HAZARD ANALYSIS

To analyze the activities involved in sampling, an Activity Hazard Analysis (AHA) may be required. An example AHA is given in Table 1.1.

1.5 SECURITY

Whenever confidential or security issue data are collected, this information must be secured. One of the most difficult issues when on-site is just what to write down or record on media. Too much information is as bad as too little information.

For the junior sampling technician, raw data should not be interpreted in the field without consultation with the scope development team. Usually this consultation will produce advice to the sampling technician; do not record that advice other than through a verbal discussion with senior staff.

Table 1.1   Air Sampling and Monitoring

Activity	Hazards	Recommended Controls
Air sampling and monitoring	Electrical	• Grounded plugs should be used. • Generators or air pumps should be used in dry areas, away from possible ignition sources. • Do not stand in water or other liquids when handling equipment. • Electrical equipment will conform to OSHA 1910.303(a) and 1910.305(a),(f),(f)(3). • Ground fault interrupters are used in the absence of properly grounded circuitry or when portable tools must be used in wet areas. • Extension cords should be protected from damage and maintained in good condition.
	Sampling pumps Ambient environment and readings	• Air pumps should be placed within easy reach using an OSHA-approved ladder or elevated platform or by placing the pump on a stake. • Personnel should be thoroughly familiar with the use, limitations, and operating characteristics of the monitoring instruments. • Perform continuous monitoring in variable atmospheres. • Use intrinsically safe instruments until the absence of combustible gases or vapor is anticipated.

Samples should be handled only by workers specifically designated as samplers. The worker who signs the chain-of-custody record will guarantee sample integrity until its final arrival at the laboratory.

1.5.1 Sample Containers—Laboratory

The analytical laboratory will often provide sample containers. The containers for soil or water sampling will be either high-density polyethylene or glass with Teflon®-lined lids and will be pretreated with preservatives as applicable. The type of sample containers and preservatives required for each analysis will be specified by the laboratory in coordination with the scope of work.

Sample filter cassettes, sorbent tubes, and other collection devices for air samples may also be obtained for the laboratory. Coordination with the laboratory is essential to ensure that sample containers meet the laboratory's internal quality control requirements as well as regulatory requirements.

1.5.2 Sample Handling and Decontamination

After sample collection in the field, the exterior of sample containers will be decontaminated if gross contamination is present. The sample containers will be handled with gloves until they are decontaminated with a detergent wash and water rinse. Care will be taken to avoid damaging the temporary labeling during decontamination. After decontamination, permanent labels will be placed on clean sample container exteriors.

The sample containers will be well cushioned with packing materials and packaged as described below for transportation to the laboratory. Care will be taken to seal bottle caps tightly. The samples will be shipped to the laboratory under chain-of-custody protocols.

Asbestos samples should never be sent with packing peanuts because the static charge generated during shipping will alter the pattern of fiber deposition on the cassette filters. Volatile samples must be sent in cooling chests to maintain a 4°C atmosphere during shipment. Semivolatiles should also be sent in cooling chests.

1.5.3 Procedures for Packing and Shipping Low Concentration Samples

Samples will be packaged as follows:

- Use waterproof metal (or equivalent strength plastic) ice chests or coolers only.
- After determining the specific samples to be submitted and filling out the pertinent information on the sample label and tag, put the label on the bottle or vial prior to packing.
- Secure the lid with strapping tape (tape on volatile organic compound [VOC] vials may cause contamination).
- Mark volume level on bottle with grease pencil.
- Place about 3 in. of inert cushioning material, such as vermiculite, in the bottom of the cooler.
- Enclose the bottles in clear plastic bags through which sample tags and labels are visible and seal the bags. Pack bottles upright in the cooler and isolate them in such a way that they *do not touch* and will not touch during shipment.
- Place bubble wrap and/or packing material around and among the sample bottles to partially cover them (no more than halfway).
- Add sufficient ice (double bagged) between and on top of the samples to cool them and keep them at approximately 4°C until received by the analytical laboratory.
- Fill cooler with cushioning material.
- Put paperwork (chain-of-custody record) in a waterproof plastic bag and tape it with duct tape to the inside.
- Tape the drain of the cooler shut with duct tape.
- Secure the lid by wrapping the cooler completely with strapping, duct, or clear shipping tape at a minimum of two locations. Do not cover any labels.
- Attach completed shipping label to the top of the cooler.
- Label "This Side Up" on the top of the cooler, "Up" with arrow denoting direction on all four sides, and "Fragile" on at least two sides.
- Affix numbered and signed custody seals on front right and back left of cooler. Cover seals with wide, clear tape.

1.5.4 Procedures for Packing and Shipping Medium Concentration Samples

An effort will be made to identify samples suspected of having elevated contaminant concentrations based on field observations and screening tests. These samples will be segregated and packed in a separate container to the extent allowed by prevailing field conditions. Lids for these samples will be sealed to the containers with tape. Medium concentration samples will be packed in the same manner as described for low concentration samples.

1.5.5 Chain-of-Custody Records

Chain-of-custody protocols will be established to provide documentation that samples were handled by authorized individuals as a means to maintain sample integrity. The chain-of-custody record will contain the following information:

- Sample identification number
- Date, time, and depth of sample collection
- Sample type (e.g., sludge)
- Type and number of container
- Requested analyses
- Field notes and laboratory notes
- Project name and location
- Name of collector
- Laboratory name and contact person
- Signature of person relinquishing or receiving samples

Chain-of-custody records will be maintained for each laboratory sample. At the end of each day on which samples are collected, and prior to the transfer of the samples off-site, chain-of-custody documentation will be completed for each sample. Information on the chain-of-custody record will be verified to ensure that the information is consistent with the information on the container labels and in the field logbook.

Upon receipt of the sample cooler at the laboratory, the laboratory custodian will break the shipping container seal, inspect the condition of the samples, and sign the chain-of-custody record to document receipt of the sample containers. Information on the chain-of-custody record will be verified to ensure that the information is consistent with the information on the container labels. If the sample containers appear to have been opened or tampered with, this discrepancy should be noted by the person receiving the samples under the section entitled "Remarks." The completed chain-of-custody records will be included with the analytical report prepared by the laboratory.

1.5.6 Mailing—Bulk and Air Samples

Mail bulk samples and air samples separately to avoid cross-contamination:

- Pack the samples securely to avoid any rattle or shock damage. Do not use expanded polystyrene packaging.
- Use bubble sheeting as packing.
- Put identifying paperwork in every package.
- Do not send samples in plastic bags or envelopes.
- Do not use polystyrene packing peanuts.
- *Print legibly on all forms.*

For exceptional sampling conditions or high flow rates, contact a Certified Industrial Hygienist (CIH) or the chosen analytical laboratory (approved by the American Industrial Hygiene Association [AIHA]).

1.6 EQUIPMENT PRECAUTIONS

1.6.1 BATTERIES

1.6.1.1 Alkaline Batteries

Replace frequently (once a month) and carry fresh replacements.

1.6.1.2 Rechargeable Nickel-Cadmium (Ni-Cad) Batteries

Check the batteries under load (e.g., turn pump on and check voltage at charging jack) before use. See the manufacturer's instructions for locations to check voltage. Use 1.3 to 1.4 V per Ni-Cad cell for an estimate of the fully charged voltage of a rechargeable battery pack.

It is undesirable to discharge a multicell Ni-Cad battery pack to voltage levels that are 70% or less of its rated voltage; this procedure will drive a reverse current through some of the cells and can permanently damage them. When the voltage of the battery pack drops to 70% of its rated value; it is considered depleted and should be recharged.

Rechargeable Ni-Cad batteries should be charged only in accordance with the manufacturer's instructions. Chargers are generally designed to charge batteries quickly (approximately 8 to 16 h) at either a high charge rate or slowly (trickle charge). A battery can be overcharged and ruined when a high charge rate is applied for too many hours. However, Ni-Cad batteries may be left on trickle charge indefinitely to maintain them at peak capacity. In this case discharging for a period equal to the longest effective field service time may be necessary, because of short-term memory imprinting (Figure 1.1).

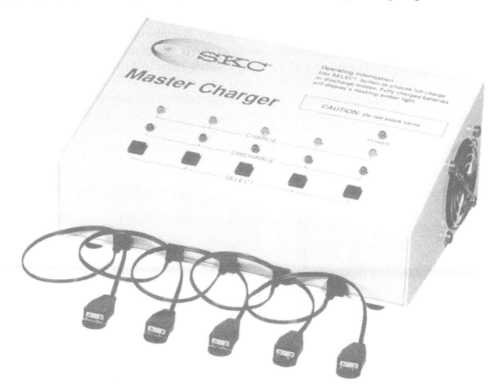

Figure 1.1   This battery maintenance system accommodates one to five rechargeable air sampling pump battery packs. (SKC)

1.7 ADVERSE TEMPERATURE EFFECTS

High ambient temperature, above 100°F and/or radiant heat (e.g., from nearby molten metal) can cause flow faults in air sampling pumps. If these conditions are likely, use the pump with a higher operating temperature range (e.g., Dupont Alpha-1) as opposed to a pump with a lower operating temperature range (e.g., SKC).

1.8 EXPLOSIVE ATMOSPHERES

Instruments must not be used in atmospheres where the potential for explosion exists (29 CFR 1910.307). Instruments must be intrinsically safe and certified by the

- Mine Safety and Health Administration (MSHA)
- Underwriter's Laboratory (UL)
- Factory Mutual (FM)
- Other testing laboratories recognized by the Occupational Safety and Health Administration (OSHA)

When batteries are being replaced, use only the type of battery specified on the safety approval label.

Do not assume that an instrument is intrinsically safe. If you are uncertain, verify its safety by contacting the instrument's manufacturer.

1.9 ATMOSPHERES CONTAINING CARCINOGENS

A plastic bag should be used to cover equipment when carcinogens are present. Decontamination procedures for special environments should be followed after using equipment in carcinogenic environments.

CHAPTER 2

Air Sampling Instrumentation Options

This chapter details and discusses the options available for monitoring various contaminants. It includes information for contaminant mixes, thermal enthalpy, interferences, and basis calibration. It also provides cross-section diagrams to illustrate the internal function of various detector and sensor elements.

2.1 VOLATILE ORGANIC COMPOUNDS

Sampling for volatile organics essentially means sampling for carbon-containing compounds that can get into the air. The term *volatile* usually means that the chemical gets into the air through a change of phase from liquid to gas. This phase change occurs when temperatures approach, equal, and exceed the boiling point and continue until equilibrium is established in the environment.

For a chemical with a boiling point over 100°F, we would not expect to find that chemical volatilizing at room temperatures. A chemical with a boiling point of 75°F, on the other hand, would be expected to readily volatilize into the environment.

Unfortunately, like so many rules, this one is not always true. Volatilization can imply that the chemical is being transported in the airstream by mechanical means that exposes surface area. An example of this anomaly is mercury, which has a boiling point of 674°F. Mercury as a liquid can be dispersed into the airstream as tiny droplets. The phase change occurs around each of these droplets as an equilibrium is established between the mercury liquid and the mercury in the immediate area gas phase. Thus mercury vapor is dispersed into the atmosphere by an equilibrium volatilization phenomenon that is more dependent on mechanical dispersion than on temperature differentials.

2.1.1 Photoionization Detector (PID)

Some volatile chemicals can be ionized using light energy. Ionization is based on the creation of electrically charged atoms or molecules and the flow of these positively charged particles toward an electrode. Photoionization (Figure 2.1) is accomplished by applying the energy from an ultraviolet (UV) lamp to a molecule to promote this ionization. A PID is an instrument that measures the total concentration of various organic vapors the in the air.

Molecules are given an ionization potential (IP) number based on the energy needed to molecularly rip them apart as ions. Chemicals normally found in the solid and liquid state

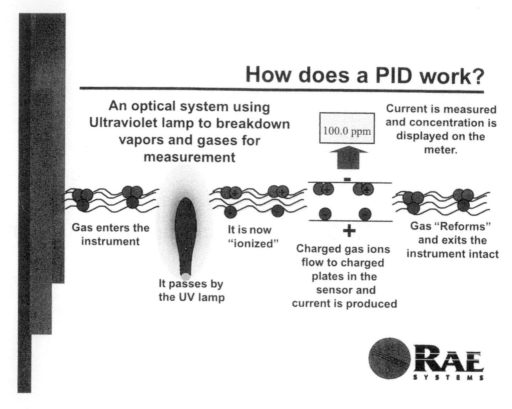

Figure 2.1   Photoionization detector working diagram. (RAE Systems)

at room temperatures do not have an IP. By definition IPs are given to chemicals found at room temperature as gases (Figure 2.2).

If the IP is higher than the energy that can be transmitted to a molecule by the UV lamp, the molecule will not break apart. Other energy sources can be used from other instruments, such as the flame ionization detector (FID) that has a hydrogen gas flame to impart energy to molecules; of course, these detectors are not called PIDs.

The PID is a screening instrument used to measure a wide variety of organic and some inorganic compounds. The PID's limit of detection for most volatile contaminants is approximately 0.1 ppm. The instrument (Figure 2.3) has a handheld probe. The specificity of the instrument depends on the sensitivity of the detector to the substance being measured, the number of interfering compounds present, and the concentration of the substance being measured relative to any interferences.

Newer PIDs have sensitivities down to the parts per billion range. These instruments utilize very high-energy ionization lamps. When toxic effects can occur at the parts per billion range, such as with chemical warfare agents or their dilute cousins—pesticides and other highly hazardous chemicals—these newer PIDs are essential (Figure 2.4).

Some PIDs are FM approved to meet the safety requirements of Class 1, Division 2, hazardous locations of the National Electrical Code.

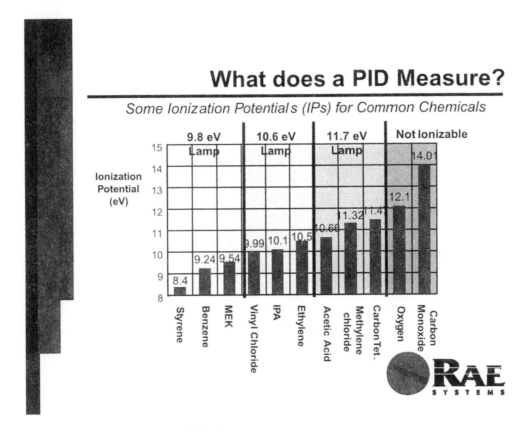

Figure 2.2   Ionization potentials. (RAE Systems)

Figure 2.3   Photoionization detector with a 10.6 eV detector. (RAE Systems)

Figure 2.4   Handheld VOC monitor with parts per billion detection. (RAE Systems)

2.1.1.1 Calibration

An instrument is calibrated by introducing pressurized gas with a known organic vapor concentration from a cylinder into the detector housing. Once the reading has stabilized, the display of the instrument is adjusted to match the known concentration. A calibration of this type is performed each day prior to using the PID (Figure 2.5).

If the output differs greatly from the known concentration of the calibration gas, the initial procedure to remedy the problem is a thorough cleaning of the instrument. The cleaning process normally removes foreign materials (i.e., dust, moisture) that affect the calibration of the instrument. If this procedure does not rectify the problem, further troubleshooting is performed until the problem is resolved. If field personnel cannot resolve the problem, the instrument is returned to the manufacturer for repair, and a replacement unit is shipped to the site immediately. The manufacturer's manual must accompany the instrument.

The PID must be kept clean for accurate operation. All connection cords used should not be wound tightly and are inspected visually for integrity before going into the field. A battery check indicator is included on the equipment and is checked prior to going into the

Figure 2.5   Calibration gases. (SKC)

field and prior to use. The batteries are fully charged each night. The PID should be packed securely and handled carefully to minimize the risk of damage.

A rapid procedure for calibration involves bringing the probe close to the calibration gas and checking the instrument reading. For precise analyses it is necessary to calibrate the instrument with the specific compound of interest. The calibration gas should be prepared in air.

2.1.1.2 Maintenance

Keeping an instrument in top operating shape means charging the battery, cleaning the UV lamp window and light source, and replacing the dust filter. The exterior of the instrument can be wiped clean with a damp cloth and mild detergent if necessary. Keep the cloth away from the sample inlet, however, and do not attempt to clean the instrument while it is connected to an electrical power source.

2.1.2 Infrared Analyzers

The infrared analyzer can be used as a screening tool for a number of gases and vapors and is presently recommended by OSHA as a screening method for substances with no feasible sampling and analytical method (Figure 2.6). These analyzers are often factory programmed to measure many gases and are also user programmable to measure other gases.

A microprocessor automatically controls the spectrometer, averages the measurement signal, and calculates absorbance values. Analysis results can be displayed either in parts

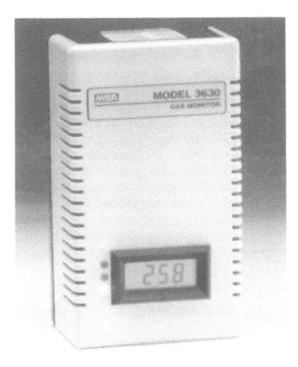

Figure 2.6 An infared gas monitor measures carbon dioxide and sends a signal to the ventilation control system.

per million or absorbance units (AU). The variable path-length gas cell gives the analyzer the capability of measuring concentration levels from below 1 ppm up to percent levels.

Some typical screening applications are as follows:

- Carbon monoxide and carbon dioxide, especially useful for indoor air assessments
- Anesthetic gases, e.g., nitrous oxide, halothane, enflurane, penthrane, and iso-flurane
- Ethylene oxide
- Fumigants, e.g., ethylene dibromide, chloropicrin, and methyl bromide

The infrared analyzer may be only semispecific for sampling some gases and vapors because of interference from other chemicals with similar absorption wavelengths.

2.1.2.1 Calibration

The analyzer and any strip-chart recorder should be calibrated before and after each use in accordance with the manufacturer's instructions.

2.1.2.2 Maintenance

No field maintenance of this device should be attempted except for items specifically detailed in the instruction book, such as filter replacement and battery charging.

2.1.3 Remote Collection

Various containers may be used to collect gases for later release into laboratory analytical chambers or sorbent beds. The remote collection devices include bags (Figure 2.7), canisters (Figure 2.8), and evacuation chambers. Remote collection refers to the practice of collecting the gas sample, hopefully intact, at a site remote from the laboratory where analysis will occur.

This method of sample collection must always take into account the potential of the collecting vehicle reacting with the gaseous component collected during the time between collection and analysis. For this reason various plastic formulations and stainless steel compartments have been devised to minimize reactions with the collected gases.

When bags are used, the fittings for the bags to the pumps must be relatively inert and are usually stainless steel (Figure 2.9). Multiple bags may be collected and then applied to a gas chromatograph (GC) column using multiple bag injector systems (Figure 2.10).

One innovation in remote sampling of this type is the MiniCan. This device can be preset to draw in a known volume of gas. The MiniCan is then worn by a worker or placed in a static location. Sample collection then occurs without the use of an additional air-sampling pump (Figure 2.11).

2.1.4 Oxygen/Combustible Gas Indicators (O₂/CGIs)/Toxin Sensors

To measure the lower explosive limit (LEL) of various gases and vapors, these instruments use a platinum element or wire as an oxidizing catalyst. The platinum element is one leg of a Wheatstone bridge circuit. These meters measure gas concentration as a percentage of the LEL of the calibrated gas (Figure 2.12).

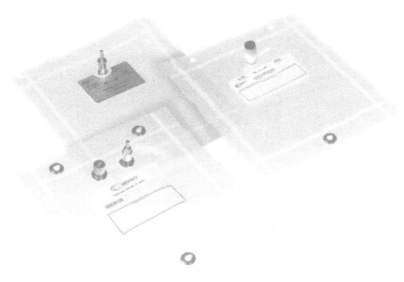

Figure 2.7 Gas sample bags are a convenient means of collecting gas and vapor samples in air. (SKC)

Figure 2.8 Six-liter canisters can be used for the passive collection of ambient VOCs from 0.1 to 100 ppb over a period of time. (SKC)

The oxygen meter displays the concentration of oxygen in percent by volume measured with a galvanic cell. Some O_2/CGIs also contain sensors to monitor toxic gases/vapors. These sensors are also electrochemical (as is the oxygen sensor). Thus, whenever the sensors are exposed to the target toxins, the sensors are activated.

Other electrochemical sensors are available to measure carbon monoxide (CO), hydrogen sulfide (H_2S), and other toxic gases. The addition of two toxin sensors, one for H_2S and one for CO, is often used to provide information about the two most likely contaminants of concern, especially within confined spaces. Since H_2S and CO are heavier than

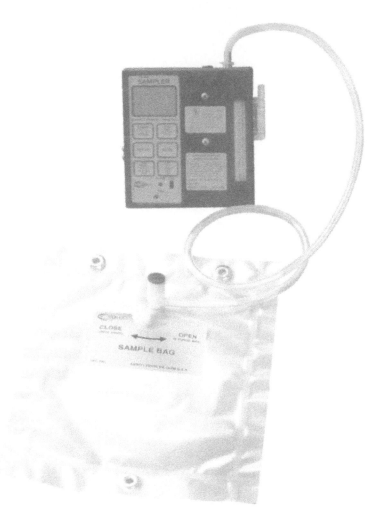

Figure 2.9   Air sampling pump connected to a Tedlar Bag. (SKC)

ambient air (i.e., the vapor pressure of H_2S is greater than one), the monitor or the monitor's probe must be lowered toward the lower surface of the space/area being monitored.

Other toxic sensors are available; all are electrochemical. Examples are sensors for ammonia, carbon dioxide, and hydrogen cyanide. These sensors may be installed for special needs.

2.1.4.1 Remote Probes and Diffusion Grids

With a remote probe, air sampling can be accomplished without lowering the entire instrument into the atmosphere. Thus, both the instrument and the person doing the sampling are protected. The remote probe has an airline (up to 50 ft) that draws sampled air toward the sensors with the assistance of a powered piggyback pump. Without this arrangement the O_2/CGI monitor relies on a diffusion grid (passive sampling).

All O_2/CGIs must be positioned so that either the diffusion grids over the sensors or the inlet port for the pumps are not obstructed. For instance, do not place the O_2/CGI on your belt with the diffusion grids facing toward your body.

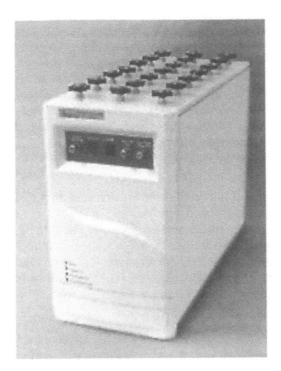

Figure 2.10 The Tedlar Bag Autosampler automates the introduction of up to 21 samples into a GC for quantitative analysis. (Entech Instruments Inc.)

Figure 2.11 Stainless steel canisters are used for collecting air samples of VOCs and sulfur compounds over a wide concentration range (1 ppb to 10,000 ppm). This 400-cc unit can be placed at a sampling site for area sampling or attached onto a worker's belt for personal sampling. (SKC-MiniCans)

2.1.4.2 Calibration Alert and Documentation

A calibration alert is available with most O_2/CGIs to ensure that the instruments cannot be used when factory calibration is needed. Fresh air calibration and sensor exposure gas calibration for LEL levels and toxins can be done in the field. However, at approximately

Figure 2.12   Multigas meters are available to allow the user to select as many as five sensors that can be used at one time. (MSA—Passport FiveStar Alarm)

6–12 month intervals, and whenever sensors are changed, factory calibration is required to ensure that electrical signaling is accurate.

Always calibrate and keep calibration logs as recommended by the manufacturer. In lieu of the manufacturer's recommendations, O_2/CGIs must be calibrated at least every 30 days.

If O_2/CGIs are transported to higher elevations (i.e., from Omaha to Denver) or if they are shipped in an unpressurized baggage compartment, recalibration may be necessary. Refer to the manufacturer's recommendations in these cases.

2.1.4.3 Alarms

Alarms must be visible and audible, with no opportunity to override the alarm command sequence once initiated and while still in the contaminated alarm-initiating environment. The alarm can be enhanced up to 150 dBA. The alarm must be wired so that the alarm signal cannot be overridden by calibration in a contaminated environment and thus cease to provide valid information.

An audible alarm that warns of low oxygen levels or malfunction or an automatic shutdown feature is very important because without adequate oxygen, the CGI will not work correctly.

2.1.4.4 Recommendations for O_2/CGIs

At a minimum, all O_2/CGIs must contain sensors for detecting levels of oxygen and the LEL percentage of the vapors/gases in the area. In an oxygen-depleted or oxygen-enriched environment, the LEL sensor will burn differently (slower in an oxygen-depleted environment and faster in an oxygen-enriched environment). Thus, in an oxygen-depleted environment the LEL sensor will be slower to reach the burn rate the monitor associates with 10% of the LEL of the calibration gas and vice versa. Consequently, all O_2/CGIs must monitor and alarm first on the basis of the oxygen level, then in response to LEL or toxin levels.

- The oxygen monitor must be set to alarm at less than 19.5% oxygen (oxygen-depleted atmosphere, hazard of asphyxiation) and greater than 22% oxygen (oxygen-enriched atmosphere, hazard of explosion/flame). **Note**: The confined space regulation for industry (29 CFR 1910.146) defines an oxygen-enriched atmosphere at greater than 23.5% oxygen.
- The LEL must be set to alarm at 10% in confined space entries.

This alarm should be both audible and visible. The alarm should not reset automatically. In other words, a separate action on the part of the user should be required to reset the alarm.

The oxygen sensor is an electrochemical sensor that will degrade as the sensor is exposed to oxygen. Thus, whether the sensor is used or not, the oxygen sensor will degrade in a period of 6 to 12 months.

Some manufacturers recommend storing the monitor in a bag placed in a refrigerated compartment. This procedure has minimal value. Because the oxygen sensor is constantly exposed to oxygen and will degrade (regardless of usage), O_2/CGIs should be used often and continuously—"there is no saving them!" In other words, once the O_2/CGI is turned on, leave the O_2/CGI on. Do not turn the monitor "on and off."

2.1.4.5 Relative Response

When using O_2/CGIs to monitor the LEL, remember that calibration to a known standard is necessary; all responses to any other gases/vapors will be relative to this calibration standard. Thus, if the O_2/CGI is calibrated to pentane (five-carbon chain), the response to methane (one-carbon chain) will be faster. In other words, less of the methane will be necessary in order for the monitor to show 10% of the LEL than if the sensor was exposed to pentane.

The LEL sensor is a platinum wire/filament located on one side of a Wheatstone bridge electrical circuit. As the wire is exposed to gases/vapors, the burn rate of the filament is altered. Thus, the resistance of the filament side of the Wheatstone bridge is changed. The O_2/CGI measures this change in resistance.

- The LEL sensor functions only when the O_2/CGI is in use; therefore, some manufacturers will state that usage of the O_2/CGI accelerates the breakdown of this sensor. However, because the oxygen sensor is much more susceptible to degradation regardless of usage, turning the monitors on and off just to preserve the LEL sensor is not recommended.
- Remember that the LEL readout is a percentage of the LEL listed for each chemical. Thus, if the LEL for a particular calibration gas is 2%, at 10% of the LEL, 0.2% of the calibration gas is present. This comparison is not possible for other than the calibrated gas/vapor atmospheres. As an example, when an O_2/CGI is calibrated to pentane and then taken into an unknown atmosphere, then at 10% of the LEL, the sensor's burn rate is the same as if the sensor had been exposed to 10% of the LEL for pentane.

If atmospheres of methane or acetylene are known to be present, the O_2/CGI must be calibrated for these gases.

2.1.4.6 Relative Response and Toxic Atmosphere Data

No direct correlation can be made under field conditions between the LEL monitor and the level of toxins. Thus, 10% (1×10^{-2}) LEL readings cannot be converted to parts per million (ppm, 1×10^{-6}) by simply multiplying by 10,000. In a controlled laboratory atmosphere using only the atmosphere to which the CGIs were calibrated, and then using many trials of differing atmospheres, relative monitoring responses and correlation to toxin levels may be obtained. However, in the field, and particularly in relatively unknown constituent atmospheres, such correlations cannot be made.

2.1.4.7 Special Considerations

- Silicone compound vapors, leaded gasoline, and sulfur compounds will cause desensitization of the combustible sensor and produce erroneous (low) readings.
- High relative humidity (90–100%) causes hydroxylation, which reduces sensitivity and causes erratic behavior, including inability to calibrate.
- Oxygen deficiency or enrichment such as in steam or inert atmospheres will cause erroneous readings for combustible gases.
- In drying ovens or unusually hot locations, solvent vapors with high boiling points may condense in the sampling lines and produce erroneous (low) readings.
- High concentrations of chlorinated hydrocarbons such as trichloroethylene or acid gases such as sulfur dioxide will depress the meter reading in the presence of a high concentration of combustible gas.
- High-molecular-weight alcohols can burn out the meter's filaments.
- If the flash point is greater than the ambient temperature, an erroneous (low) concentration is indicated. If the closed vessel is then heated by welding or cutting, the vapors will increase, and the atmosphere may become explosive.
- For gases and vapors other than those for which a device was calibrated, users should consult the manufacturer's instructions and correction curves.

2.1.4.8 Calibration

Before using the monitor each day, calibrate the instrument to a known concentration of combustible gas (usually methane) equivalent to 25–50% LEL full-scale concentration. The monitor must be calibrated to the altitude at which it is used. Changes in total atmospheric pressure due to changes in altitude will influence the instrument's measurement of the air's oxygen content. The instrument must measure both the level of oxygen in the atmosphere and the level a combustible gas reaches before igniting; consequently, the calibration of the instrument is a two-step process.

1. The oxygen portion of the instrument is calibrated by placing the meter in normal atmospheric air and determining that the oxygen meter reads exactly 20.8% oxygen. This calibration should be done once per day when the instrument is in use.
2. The CGI is calibrated to pentane to indicate directly the percentage LEL of pentane in air. The CGI detector is also calibrated daily when used during sampling events and whenever the detector filament is replaced. The calibration kit included with the CGI contains a calibration gas cylinder, a flow control, and an adapter hose.

The unit's instruction manual provides additional details on sensor calibration.

2.1.4.9 Maintenance

The instrument requires no short-term maintenance other than regular calibration and battery recharging. Use a soft cloth to wipe dirt, oil, moisture, or foreign material from the instrument. Check the bridge sensors periodically, at least every 6 months, for proper functioning.

2.1.5 Oxygen Meters

Oxygen-measuring devices can include coulometric and fluorescence measurement, paramagnetic analysis, and polarographic methods.

2.1.6 Solid Sorbent Tubes

Organic vapors and gases may be collected on activated charcoal, silica gel, or other adsorption tubes using low-flow pumps. Tubes may be furnished with either caps or flame-sealed glass ends. If using the capped version, simply uncap during the sampling period and recap at the end of the sampling period.

Multiple tubes can be collected using one pump. Flow regulation for each tube is accomplished using critical orifices and valved regulation of airflow. Calibration of parallel or y-connected multiple tube drawing stations must be done individually for each tube, even in cases where the pump is drawing air through more than one tube in a parallel series (Figure 2.13). In instances where tubes are connected in series, only one calibration draw is done through the conjoined tubes that empty air, one directly into the other (Figure 2.14).

Sorbent tubes may be used just to collect gases and vapors or to both collect and react with the collected chemicals. Some of the reactions may produce chemicals that when off-gassed could harm the pumps being used to pull air through the sorbent media bed. In these cases either filters or intermediate traps must be used to protect the pumps (Figure 2.15). The following protocols should be followed:

- Immediately before sampling, break off the ends of the flame-sealed tube to provide an opening approximately half the internal diameter of the tube. Take care when breaking these tubes—shattering may occur. A tube-breaking device that shields the sampler should be used.
- Wear eye protection.
- Use tube holders, if available, to minimize the hazards of broken glass (Figure 2.16).
- Do not use the charging inlet or the exhaust outlet of the pump to break the ends of the tubes.
- Use the smaller section of the tube as a backup and position it near the sampling pump.
- The tube should be held or attached in an approximately vertical position with the inlet either up or down during sampling (Figure 2.17).
- Draw the air to be sampled directly into the inlet of the tube. This air is not to be passed through any hose or tubing before entering the tube. A short length of protective tape, a tube holder, or a short length of tubing should be placed on the cut tube end to protect the worker from the jagged glass edges.
- Cap the tube with the supplied plastic caps immediately after sampling and seal as soon as possible.
- Do not ship the tubes with bulk material.

For organic vapors and gases, low-flow pumps are required. With sorbent tubes, flow rates may have to be lowered or smaller air volumes (half the maximum) used when there is high humidity (above 90%) in the sampling area or when relatively high concentrations of other organic vapors are present.

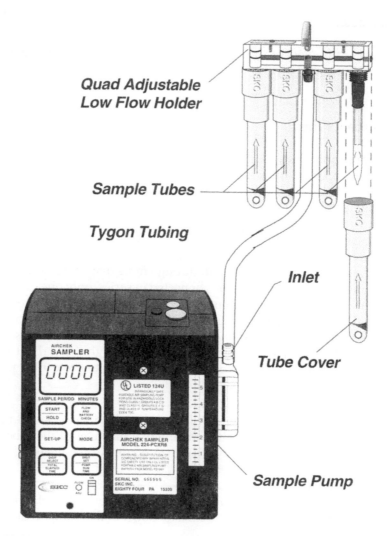

Quad Adjustable
Low Flow Holder

Sample Tubes

Tygon Tubing

Inlet

Tube Cover

Sample Pump

Figure 2.13   Multitube sampling allows sampling of multiple contaminants requiring different sampling tubes with one pump. Multitube sampling also allows you to collect time-weighted averages (TWAs) and short-term exposure limits (STELs) side by side. (SKC)

2.1.6.1 Calibration Procedures

Set up the calibration apparatus as shown in Figure 2.18, replacing the cassette and cyclone with the solid sorbent tube to be used in the sampling (e.g., charcoal, silica gel, other sorbent media). If a sampling protocol requires the use of two sorbent tubes, the calibration train must include these two tubes. The airflow must be in the direction of the arrow on the tube (Figure 2.19). Sorbent tubes may be difficult to calibrate, especially if flow-restrictive devices must be used (Figure 2.20).

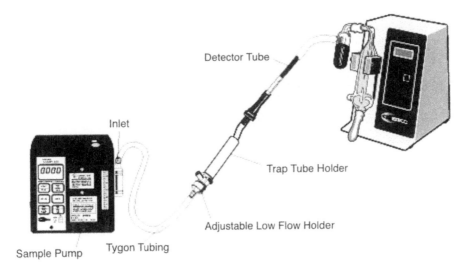

Figure 2.14 Pump with detector tube sampling train with calibrator. (SKC—pump, low-flow holder, trap tube holder, and electronic calibrator)

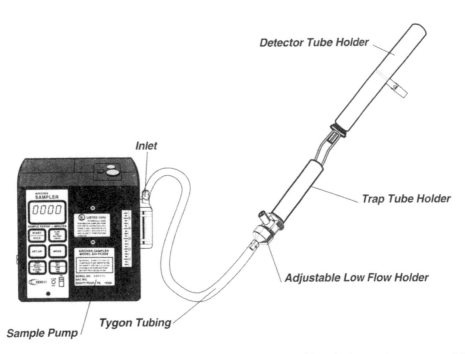

Figure 2.15 Pump with detector tube sampling train in place. Chemicals may be generated that, if allowed to enter the sampler, could damage the sampler. Therefore, a trap tube must be used between the detector tube and the sampler. (SKC—pump, low-flow holder, and trap tube holder)

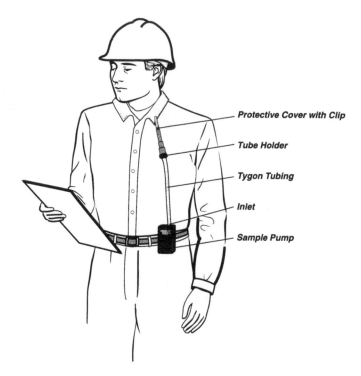

Protective Cover with Clip

Tube Holder

Tygon Tubing

Inlet

Sample Pump

Figure 2.16   Worker wearing sampling pump with sampling train in place in breathing zone. (SKC—
210 Series Pocket Pump®, low flow tube holder)

2.1.7 Vapor Badges

Passive-diffusion sorbent badges are useful for screening and monitoring certain
chemical exposures, especially vapors and gases (Figure 2.21). Badges are available from
the laboratory to detect mercury, nitrous oxide, ethylene oxide, and formaldehyde (Figure
2.22). Interfering substances should be noted.

A variation on the vapor badge is available as a dermal patch (Figure 2.23). These dermal patches can also be used in the detection of semivolatiles.

2.1.8 Detector Tubes

Detector tubes use the same medium base—silica gel or activated charcoal—as do
many sorbent tubes. The difference is that the detector tubes change color in accordance
with the amount of chemical reaction occurring within the medium base. The medium base
has been treated with a chemical that effects a given color change when certain chemicals
are introduced into the tube and reside for a time in the medium. The residence time for the
reaction to occur and the volume of air that must be drawn through the detector tubes
varies with the chemical and anticipated concentration. All detector tube manufacturers
supply the recipe for using their detector tubes as an insert sheet with the tubes.

Figure 2.17 Sorbent tube placement with protective tube holder. (SKC)

Detector tube pumps are portable equipment that, when used with a variety of commercially available detector tubes, are capable of measuring the concentrations of a wide variety of compounds in industrial atmospheres. Each pump should be leak-tested before use. Calibrate the detector tube pump for proper volume at least quarterly or after 100 tubes.

Operation consists of using the pump to draw a known volume of air through a detector tube designed to measure the concentration of the substance of interest. The concentration is then determined by a colorimetric change of an indicator that is present in the tube contents (Figure 2.24).

Most detector tubes can be obtained locally. Draeger or Sensidyne tubes are specified by some employers; their concentration detection ranges match employers' needs.

Detector tubes and pumps are screening instruments that may be used to measure more than 200 organic and inorganic gases and vapors or for leak detection. Some aerosols can also be measured.

Detector tubes of a given brand should be used only with a pump of the same brand. The tubes are calibrated specifically for the same brand of pump and may give erroneous results if used with a pump of another brand.

Cassette & Cyclone Use

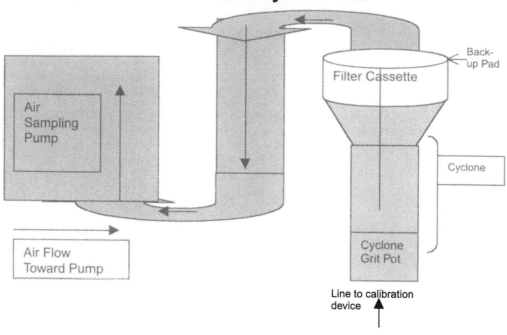

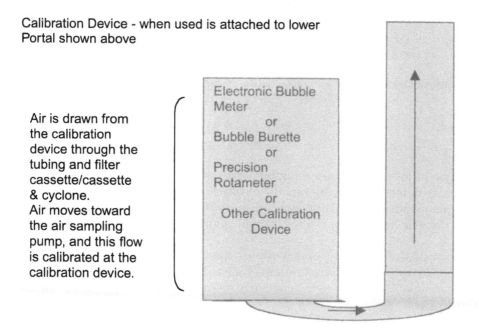

Figure 2.18 Cassette and cyclone use.

A limitation of many detector tubes is the lack of specificity. Many indicators are not highly selective and can cross-react with other compounds. Manufacturers' manuals describe the effects of interfering contaminants.

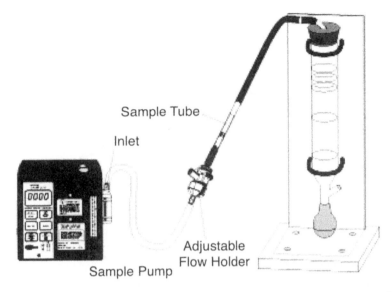

Figure 2.19 Tube sampling train connected to a sample pump and a flowmeter. (SKC—PCXR8 Sampler and Film Flowmeter)

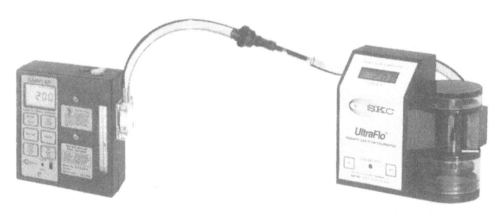

Figure 2.20 Electronic flowmeter connected to sorbent tube sampling train. (SKC—Model 709 Flowmeter)

Another important consideration is sampling time. Detector tubes give only an instantaneous interpretation of environmental hazards, which may be beneficial in potentially dangerous situations or when ceiling exposure determinations are sufficient. When long-term assessment of occupational environments is necessary, short-term detector-tube measurements may not reflect TWA levels of the hazardous substances present.

Detector tubes normally have a shelf life at 25°C of 1 to 2 years. Refrigeration during storage lengthens the shelf life. Outdated detector tubes (i.e., beyond the printed expiration date) should never be used.

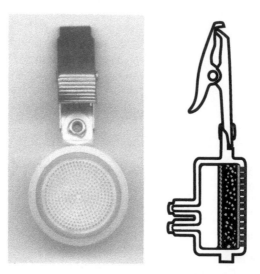

Figure 2.21   Cross-sectional view of a passive sampler. A diffusion barrier maintains sample uptake by molecular diffusion independent of wind velocity. (SKC—575 Series Passive Sampler)

Figure 2.22   Passive badge sampler. (SKC—Formaldehyde Passive Sampler)

Figure 2.23   Dermal polyurethane foam (PUF) patches for chlorinated or organonitrogen herbicides. The dermal patches are clipped onto a worker's clothing or taped to the skin in various locations where absorption may occur. After sampling, the patches are transferred to glass jars, desorbed with isopropanol, and analyzed by gas chromatography/electron capture detection (GC/ECD). (SKC)

Figure 2.24 Sorbent tube of detector tube. Flow is toward the air sampling pump in the direction of the arrow.

2.1.8.1 Performance Data

Specific models of detector tubes can be obtained from the manufacturer (e.g., Draeger, Sensidyne). The specific tubes listed are designed to cover a concentration range that is near the permissible exposure limit (PEL). Concentration ranges are tube dependent and can be anywhere from one hundredth to several thousand parts per million. The limits of detection depend on the particular detector tube. Accuracy ranges vary with each detector tube.

The pump may be handheld during operation (weight about 8–11 oz), or it may be an automatic type (weight about 4 lb) that collects a sample using a preset number of pump strokes. A full pump stroke for either type of short-term pump has a volume of about 100 ml.

In most cases where only one pump stroke is required, sampling time is about 1 min. Determinations for when more pump strokes are required take proportionately longer.

Multiple tubes can be used with newer microcapillary detector tube instruments. Computer chips are programmed to draw preselected air volumes across these detector tubes. Readout is measured based on changes in light absorption across the microcapillary tubes.

2.1.8.2 Leakage Test

Each day prior to use, perform a pump leakage test by inserting an unopened detector tube into the pump and attempt to draw in 100 ml of air. After a few minutes check for pump leakage by examining pump compression for bellows-type pumps or return to resting position for piston-type pumps. Automatic pumps should be tested according to the manufacturer's instructions.

In the event of leakage that cannot be repaired in the field, send the pump to the manufacturer for repair. Record that the leakage test was made on a direct-reading data form in the field logbook.

2.1.8.3 Calibration Test

Calibrate the detector tube pump for proper volume measurement at least quarterly. Simply connect the pump directly to the bubble meter with a detector tube in-line. Use a detector tube and pump from the same manufacturer. Wet the inside of the 100-ml bubble meter with soap solution. When performing volume calibration, experiment to get the soap bubble even with the 0 ml mark of the burette.

For *piston-type* pumps pull the pump handle all the way out (full pump stroke). Note where the soap bubble stops. For *bellows-type* pumps compress the bellows fully. For *automatic pumps* program the pump to take a full pump stroke.

For either type pump, the bubble should stop between the 95-ml and 105-ml marks. Allow 4 min for the pump to draw the full amount of air. (This time interval varies with the type of detector tube being used in-line with the calibration setup.)

Also check the volume for 50 ml (one half pump stroke) and 25 ml (one quarter pump stroke) if pertinent.

- A variance of ±5% error is permissible.
- If the error is greater than ±5%, send the pump for repair and recalibration.

Record the calibration information required on the calibration log. It may be necessary to clean or replace the rubber bung or tube holder if a large number of tubes have been taken with any pump.

2.1.8.4 Special Considerations

Detector tubes should be refrigerated when not in use to prolong shelf life. Detector tubes should not be used when they are cold. They should be kept at room temperature or in a shirt pocket for 1 h prior to use. Lubrication of the piston pump may be required if volume error is greater than 5%.

Draeger, Model 31 (Bellows)

When checking the pump for leaks with an unopened tube, the bellows should not be completely expanded after 10 min.

Draeger, Quantimeter 1000, Model 1 (Automatic)

A battery pack is an integral part of this pump.

- The pack must be charged prior to initial use.
- One charge is good for 1000 pump strokes.
- During heavy use, it should be recharged daily.

If a "U" (undervoltage) message is continuously displayed in the readout window of this pump, the battery pack should be immediately recharged.

Matheson-Kitagawa, Model 8014-400a (Piston)

When checking the pump for leaks with an unopened tube, the pump handle should be pulled back to the 100-ml mark and locked.

- After 2 min, the handle should be released carefully.
- The handle should return to a point <6 mm from zero or resting position.

After taking 100–200 samples, the pump should be cleaned and relubricated. This procedure involves removing the piston from the cylinder, removing the inlet and pressure-relief valve from the front end of the pump, cleaning, and relubricating.

Mine Safety Appliances, Samplair Pump, Model A, Part No. 46399 (Piston)

The pump contains a flow-rate control orifice protected by a plastic filter that periodically needs to be cleaned or replaced. To check the flow rate, the pump is connected to a burette, and the piston is withdrawn to the 100-ml position with no tube in the tube holder.

- After 24–26 s, 80 ml of air should be admitted to the pump.
- Every 6 months the piston should be relubricated with the oil provided.

Mine Safety Appliances Kwik-Draw Sampling Pump, Part No. 487500 (Bellows)

The pump contains a filter disk that needs periodic cleaning or replacement. The bellows shaft can be cleaned and lubricated with automotive wax if operation becomes jerky.

Sensidyne-Gastec, Model 800, Part No. 7010657-1 (Piston)

This pump can be checked for leaks as mentioned for the Kitagawa pump; however, the handle should be released after 1 min.

Periodic relubrication of the pump head, the piston gasket, and the piston check valve is needed and is use dependent.

A variation on the detector tube technology is the use of sorbent packed tubes that change color in response to ambient airflow. The application of reactive adsorbing and/or absorbing chemicals onto test strips is also used to provide a general indication of airborne contaminant levels. An example is the ozone test strip used to monitor both outdoor and indoor ozone levels (Figure 2.25).

2.1.9 Formaldehyde

Formaldehyde sampling can be accomplished by both passive and active (use of a pump) techniques.

- When long duration sampling is required in indoor air investigations, passive-sampling may be the method of choice (Figure 2.26).
- Vapor badges can be used to monitor personnel exposures.

Figure 2.25 Ozone strips provide quick indication of ambient levels of ozone in both indoor and outdoor air. Ozone strips are chemically treated to react with ozone. Test strips are placed in the area to be tested. After 10 min, compare the test strip with the color scale on the test strip package. Results display in four distinct colors from light yellow to brown. Each represents a certain level of ozone concentration. (SKC)

Figure 2.26   A formaldehyde passive air sampler for indoor air sampling. Easy to use, it is designed for long-term measurement (5 to 7 days). Its detection limit is 0.01 ppm. (SKC)

Neither of these methods is recommended for acute exposure scenarios because the sampling medium will quickly become overloaded. In acute exposure scenarios sampling with a sorbent tube attached to an air sampling pump, or a detector tube attached to a pump/bellows, is recommended. Attachment implies that the pump will be used to draw a known volume of air quickly into the media. This air will be at a concentration antici- pated to provide information, but below that which would overload the media.

2.2 OZONE METER

The ozone meter detector uses a thin-film semiconductor sensor. A thin-film platinum heater is formed on one side of an alumina substrate. A thin-film platinum electrode is formed on the other side, and a thin-film semiconductor is formed over the platinum elec- trode by vapor deposition. The semiconductor film, when kept at a high temperature by the heater, will vary in resistance due to the absorption and decomposition of ozone. The change in resistance is converted to a change of voltage by the constant-current circuit.

The measuring range of the instrument is 0.01–9.5 ppm ozone in air. The readings are displayed on a liquid crystal display that reads ozone concentrations directly. The temper- ature range is 0–40°C, and the relative humidity (RH) range is 10–80%.

The instrument is not intrinsically safe.

- The instrument must not be exposed to water, rain, high humidity, high temper- ature, or extreme temperature fluctuation.
- The instrument must not be used or stored in an atmosphere containing silicon compounds, or the sensor will be poisoned.
- The instrument is not to be used for detecting gases other than ozone. Measurements must not be performed when the presence of organic solvents, reducing gases (such as nitrogen monoxide, etc.), or smoke is suspected; readings may be low.

2.2.1 Calibration

Calibrate the instrument before and after each use. Be sure to use a well-ventilated area; ozone levels may exceed the PEL for short periods. Calibration requires a source of ozone. Controlled ozone concentrations are difficult to generate in the field, and this calibration is normally performed at the laboratory. Gas that is either specially desiccated or humidified must not be used for preparing calibration standards, as readings will be inaccurate.

2.2.2 Maintenance

The intake-filter unit-Teflon sampling tube should be clean, connected firmly, and checked before each operation. Check pump aspiration and sensitivity before each operation.

2.3 TOXIC GAS METERS

Toxic gas meters use an electrochemical voltametric sensor or polarographic cell to provide continuous analyses and electronic recording. Sample gas is drawn through the sensor and absorbed on an electrocatalytic-sensing electrode, after passing through a diffusion medium. An electrochemical reaction generates an electric current directly proportional to the gas concentration. The sample concentration is displayed directly in parts per million. The method of analysis is not absolute; therefore, prior calibration against a known standard is required. Exhaustive tests have shown the method to be linear; thus calibration at a single concentration is sufficient.

Sensors are available for sulfur dioxide, hydrogen cyanide, hydrogen chloride, hydrazine, carbon monoxide, hydrogen sulfide, nitrogen oxides, chlorine, ethylene oxide, and formaldehyde. These sensors can be combined with O_2/CGIs in one instrument (Figure 2.27).

Interference from other gases may be a problem. The sensor manufacturer's literature must be consulted when mixtures of gases are tested.

2.3.1 Calibration

Calibrate the direct-reading gas monitor before and after each use in accordance with the manufacturer's instructions and with the appropriate calibration gas.

- When calibrating under external pressure, the pump must be disconnected from the sensor to avoid sensor damage. If the span gas is directly fed into the instrument from a regulated pressurized cylinder, the flow rate should be set to match the normal sampling rate.

Figure 2.27 MSA Passport.

• Due to the high reaction rate of the gas in the sensor, substantially lower flow rates result in lower readings. This high reaction rate makes rapid fall time possible—simply by shutting off the pump. Calibration from a sample bag connected to the instrument is the preferred method.

2.4 SEMIVOLATILE ORGANIC COMPOUNDS (SVOCs)

Semivolatiles like polycyclic biphenyls (PCBs), polynuclear aromatic hydrocarbons (PAHs), dioxins, furans, and pesticides present unique sampling challenges. The term *semivolatile* is used for chemicals that do not normally volatilize into the gaseous state at room temperature (75°F; Figure 2.28).

These chemicals can enter the airstream through a variety of mechanisms, the most prevalent being dispersed as adsorbed or absorbed particulate contaminants. Heating phenomena such as smoking, direct heating of semivolatiles, and chemical usage in which semivolatiles must be applied in a heated state (asphalting) may also change the semivolatiles into volatiles.

Often very high collection rates (10–30 l/min) must be used to pick up semivolatile contaminants from the airstream (Figure 2.29). Certain EPA methods and stack sampling requirements also call for the use of very high flow rates. In these instances special pumps must be used.

2.4.1 PAHs

PAHs have K_{oc} values that are characteristic of chemicals that tend to readily adsorb to the soil particulate or to any other particulate present. PAHs are expected to bind strongly to soil and to not leach extensively to groundwater through volatilization.

Photolysis and hydrolysis do not appear to be significant PAH breakdown processes in soil. However, while little volatilization will occur from the soil, leaching to groundwater is possible. PAHs released to the water will dissolve at ambient pH. The dissociated form will degrade (hours to days).

Figure 2.28   High volume PUF tube for pesticides and PAHs. (SKC)

Figure 2.29 Dual-diaphragm pump for indoor and outdoor collection of particulates, PAHs, and other compounds requiring flows from 10 to 30 l/min. High-flow pumps are used for asbestos, PAHs in indoor air, PM10 and PM2.5 in indoor air, bioaerosol sampling, stack sampling, fenceline monitoring, and background monitoring. (SKC—AirCheck HV30 Environmental Air Sampler)

Photolysis is expected to occur near the water surface, and biodegradation in the water column is expected. Biodegradation probably becomes significant after acclimation (may take several weeks). PAHs with four or fewer aromatic rings are degraded by microbes. Transport of PAH biodegradation products to groundwater has been documented in some cases.

The mechanism for particulate dispersion first requires that the semivolatile chemical bind to a particulate. When that particulate is dispersed into the air, the semivolatile chemical is also dispersed. So the methods used for particulate sampling are also applicable for semivolatile sampling. The toxicological problems with the semivolatile chemical on particulate dispersion come into play as the particulate is inhaled, and off-gassing occurs in the body of the semivolatile chemical. The following decision logic is an example of the evaluation of semivolatile exposure potential, in this case PAHs, at outdoor sites:

- Usually 5 mg/m³ is assumed to be the airborne particulate concentration necessary to have visible dust. Therefore, if during sampling activities, dust is apparent, semivolatile exposure should be of concern if the semivolatile exposure limits are less than the visible dust limits.

—PAH dust exposure PEL levels are below 5 mg/m³; consequently, air monitoring would need to be conducted on any site where PAHs are expected and visible dusts are generated.

—This air monitoring should be downwind from activities judged invasive of soil layers and potentially subject to dust cloud generation.

- Monitoring for PAH levels is conducted using a personal air-sampling pump. The exposure target limit is usually set at 0.2 mg/m³ because that is the limit for Benzo(a)pyre. The setting of target limits for semivolatiles is usually based on the most toxic of the expected cogenitors. Cogenitors are essentially the different molecular conformations that a semivolatile chemical can take.

- Complete suspension of these contaminants within an airstream on-site is not physically probable, and misting of the sampling area should continually remove particulates from the air; therefore, high efficiency particulate air (HEPA) cartridge air-purifying respirators should be sufficient to protect samplers.

- Because PAH contamination is often associated with the presence of petroleum contamination, VOC levels should be continually monitored using a PID. If the PID records a sustained deflection of 1 ppm, workers should evacuate the exclusion zone. The presence of volatile organics revealed by sustained PID readings will require further site evaluation for PAH potential exposure.

—Further evaluation for potential exposures to PAHs will require soil sampling, with attendant air dispersion calculations and air monitoring for particulates.

—Unfortunately, we do not have real-time instrumentation to monitor for PAHs. PAH sampling requires that laboratory analyticals or on-site immunoassay testing must be accomplished. Therefore, until results are obtained and interpreted, on-site personnel would be required to wear HEPA-OV cartridge full-face air-purifying respirators.

The on-site monitoring sequence is as follows:

- Visible dust: 2 l/min personal air-sampling pumps will be used to draw air through filter cassettes. Cassettes will be packaged and sent to the contract laboratory for analysis.
- Ongoing site work will continue with dust suppression engineering controls. Personnel will don HEPA cartridge air-purifying respirators.
- If organic vapors are also present, HEPA-OV combination cartridges should be donned.

The above example illustrates the need for careful evaluation of the potential for exposure during sampling. Monitoring decisions when semivolatiles are present must always be made by personnel who understand that these chemicals cannot be detected by the sense of smell or predicted by visible dust.

2.4.2 PCBs and Creosote

PCBs are expected in electrical transformers, fluorescent light fixture ballasts, and possibly on some building materials where oils have leaked or the transformers have exploded. Generally, all transformers and suspect containers are inspected to determine the contents of the liquids present. Site inspections include a review of transformer labels or other identifying signage. This review is conducted in the field by comparing the reviewed information against known PCB-containing placards or by conducting a phone conversation with the owner of the transformer or the manufacturer. When containers suspected of containing PCBs are found, representative samples may be collected to evaluate the content and concentrations. These samples are evaluated in the field by utilizing Rapid Assay Analysis (RAA) for PCBs or through laboratory analysis.

Building materials suspected of being contaminated with PCBs, such as concrete beneath a leaking transformer, may be sampled and analyzed following the same procedures. In the case of stained concrete, a sample would be obtained by drilling the concrete and preparing the dust for analysis.

RAA is a method utilizing amino assay techniques for field specific analysis. The amino assay field test kit is prepared in the laboratory for analyzing a specific compound, in this case PCBs or Aroclors.

2.4.3 Pesticides and PAHs—PUF Tubes

Both pesticides and PAHs can be collected in PUF tubes. PUF tubes are available for both high-volume and low-volume sampling (Figure 2.30). The sampling volume requirement is determined by the regulatory onus and the chemical constituency of the anticipated sample.

Figure 2.30 Low volume PUF tubes for pesticides (for EPA Methods TO-10A and IP-8 and ASTM D4861 and D4947) designed for sampling common pesticides, including organochlorine, organophosphorus, pyrethrin, triazine, carbonate, and urea. (SKC)

2.5 ACID GASES OR CAUSTICS

Volatile acid gases may be an inappropriate designation. Acid gases are often generated during a reaction, and the latent volatility of the acid gas is really not an issue. Thermal volatilization based on boiling point predictions and mechanical dispersion may be of less importance than the rate of the reaction generating the acid gas or caustic. However, in addition to this reaction phenomenon, acid gases such as chlorine are given off when the liquid solution is distributed around an area. Here we have a classic case of the liquid to gas interface seeking an equilibrium. If air currents sweep the generated gas concentration away from this equilibrium site, the liquid will again yield molecules to the gas phase to again achieve another equilibrium.

Acid gases and caustics with their corrosive or caustic properties can have health effects that include both acute toxicological and physical manifestations, such as watering eyes and respiratory tract irritation. Because of these effects, sampling for acid gases and caustics must begin upon approach to the area of concern.

Sampling for acid gases and caustics may use all of the techniques specified for any volatile. Some acid gases and caustics are dispersed and adsorbed to particulate; therefore, particulate sampling techniques apply.

The reaction phenomenon must always be considered during any sampling of acid gases or caustics. Any real-time instrumentation with unprotected metal sensors, lamp filaments, or sensor housings will often be rendered useless, as the acid gases or caustics interact with the metals through reduction-oxidation (redox) reactions.

2.5.1 Impingers

Impingers may be used to bring acid/caustic-laced particulates into solutions that are retained within the impinger's vessel. Vapors, mists, and gases may also be introduced into the impinger solution. When the reaction within the impinger vessel may cause off-gassing, a filter or media barrier (see Figures 2.31, 2.32, and 2.33) may be required between the air sampling pump and the impinger vessel tubing to the pump (Figure 2.34).

Midget impingers may be worn as personal sampling devices (Figure 2.35). The main concern with impingers as sampling devices, especially for personnel, is the danger of spills.

2.5.2 Sorbent Tubes

Sampling media must also be acid and caustic resistant. Sampling for acids and caustics is often discussed in terms of using silica gel sorbent tubes. The procedure for this sampling is the same as that for volatiles where charcoal tubes are often used. The essential problem with the silica gel tubes is that they tend to plug up! The use of dual flow tubes is some insurance that if one tube plugs up, the other might still remain effective to provide data from the sampling interval.

In instances where silica gel tubes continue to *plug up,* switching to larger bore silica gel tubes or altering the sampling interval (less time) may be needed. If this procedure does not work, switching to charcoal tubes may be the only other solvent tube option. These sampling routines (see Figure 2.36) may be at odds with the recommended National Institute of Occupational Safety and Health (NIOSH) methods that may call for small bore silica gel tubes at low flow rates for extended periods of time. If so, decision logic must be documented, with this documentation linked to the competency of the individual who devised the sampling plan.

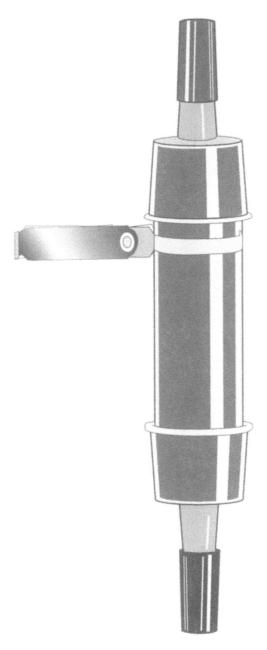

Figure 2.31 In-line traps with replacement sorbent tubes are connected between the pump and impinger holder to protect the pump. (SKC)

2.5.3 Detectors

Various detector tubes are available for acid gases and caustics. Chemical-specific detectors are increasingly available as hard-wired permanent detectors based on electro-*chemical sensors*. As with any other electrochemical sensor, recovery of the sensor after overdosing with a chemical may take time or may not be possible at all.

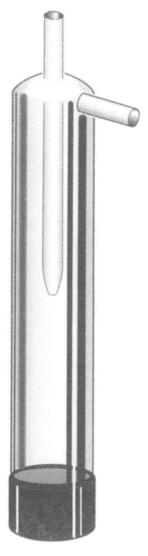

Figure 2.32   Impinger trap to prevent impinger liquids from being drawn into the sample pump. Solid
sorbents may be added to the trap when volatile liquids are used to protect the pump
chambers from exposure to vapors. (SKC)

2.5.4 pH Litmus Paper or Meter

pH litmus papers or meters are particularly valuable on sites where acid gases and
caustics may be of concern. Many sampling events require concurrent bulk sampling, and
often the pH of these samples can be effectively characterized in the field.

2.5.4.1 Calibration

Calibrations for pH meters generally follow this regime:

- Temperature and conductance are factory calibrated.
- To recalibrate conductance in the field (if necessary):

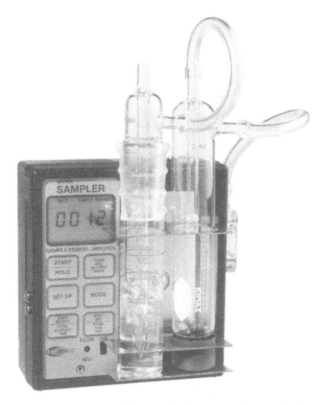

Figure 2.33   Impinger and in-line trap holder mounted on sample pump. Use a trap impinger to prevent impinger liquids from being drawn into the sample pump. Solid sorbents may be added to the trap to protect the pump chamber from exposure to vapors when volatile liquids are used. (SKC)

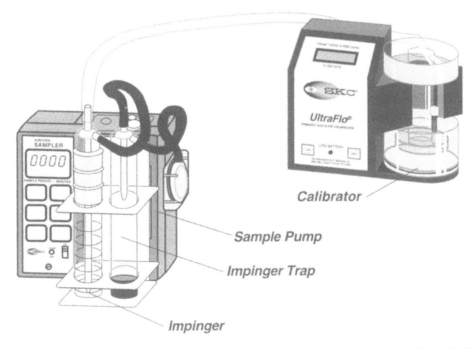

Figure 2.34   Impinger/trap sampling train with flowmeter. (SKC—Universal Sampler with Double Impinger Holder attached to UltraFlo® electronic calibrator)

Figure 2.35 Teflon PFA (fluoropolymer) impingers. These vessels are completely inert to virtually all chemicals and perform well in both high temperature and cryogenic applications. (SKC)

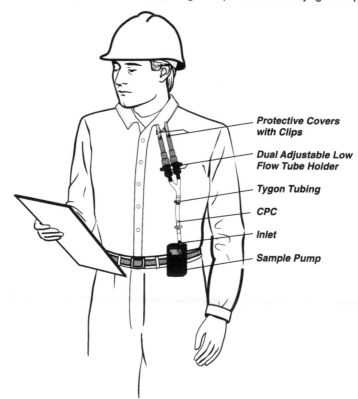

Figure 2.36 Worker wearing sampling pump and two tubes side-by-side for simultaneous tube sampling. (SKC)

—Remove the black plug revealing the adjustment potentiometer screw.
—Add standard solution to a cup, discard, and refill.
—Repeat the detection procedure until the digital display indicates the same value twice in a row.
—Adjust the potentiometer until the digital display indicates the known value of conductance.
—Increase the digital display reading by turning the adjustment potentiometer screw counterclockwise.
—Decrease the digital display reading by turning the adjustment potentiometer screw clockwise.

To standardize the pH electrode:

* Place the pH electrode in the 7.0 buffer bottle.
* Adjust the "zero" potentiometer on the face of the tester to "7.00."
* Place the pH electrode in the 4.0/10.0 buffer bottle (depending on range of concern at your site).
* Adjust the "slope" potentiometer on the face of the tester to either 4.0 or 10.0.
* Repeat the "zero" and "slope" adjustments several times—to ensure interaction stability.

2.6 MERCURY ANALYZER/GOLD FILM ANALYZER

A gold-film analyzer draws a precise volume of air over a gold-film sensor. A microprocessor computes the concentration of mercury in milligrams per cubic meter and displays the results on the digital meter.

This meter is selective for mercury and eliminates interference from water vapor, sulfur dioxide, aromatic hydrocarbons, and particulates. In high concentrations of mercury vapor the gold film saturates quickly and should not be used for concentrations expected to be over 1.5 mg/m^3. Hydrogen sulfide is an interferant. Lead may also be an interferant.

2.6.1 Jerome Mercury Analyzer

A similar instrument is manufactured by Baccarach. The Jerome is discussed here to illustrate the general principles of operation in the detection of mercury vapors.

The Jerome 431 Gold Film Sensor is inherently stable and does not require frequent calibration. The interval between calibrations is recommended at every 12 months. The Jerome Mercury Analyzer is factory calibrated using National Bureau of Standards (NBS) traceable permeation tubes. These permeation tubes have been rated at an accuracy of ±2%. Calibration includes stability of the calibration gas source being assured, elimination of any pressure differential in the calibration gas stream, and precise control of ambient temperature. Hence, calibration cannot be done in the field.

The Jerome Mercury Analyzer, equipped with a Gold Film Sensor, has these qualities:

* Rapid response time (≤4 s)
* LCD display—direct reading
* Data logger and software
* 0.003 mg/m^3 Hg sensitivity
* Accuracy ±5% at 0.107 mg/Hg

2.6.2 Survey Procedures

- Document that the instrument is calibrated.
- Obtain and record a background reading.
- Survey with the Jerome Mercury Analyzer:
  —Insert the probe in or near the area to be surveyed.
  —For surface areas hold the probe 6 in. or closer to the survey point.
  —For sink traps do not put the probe in water; allow a 10-s residence time of
     the probe in the headspace of the trap prior to sample readout.
- Document readings in the field logbook.
- Photograph general locations and specific areas of concern.

2.6.3 Precautions for Area Surveys

- Include all suspected rooms, hallways, adjacent administrative space, and storage rooms, including behind and underneath cabinets, refrigerators, sinks, and
  equipment.
- Include all locations where mercury was used or stored.
- Include all cracks, crevices, and delaminated surface areas.

2.6.3.1 Calibration

Calibration should be performed by the manufacturer or a laboratory with special
facilities to generate known concentrations of mercury vapor. Instruments should be
returned to the manufacturer or a calibration laboratory on a regular schedule.

2.6.3.2 Maintenance

Mercury vapor instruments generally contain rechargeable battery packs, filter
medium, pumps, and valves that require periodic maintenance. Except for routine charging of the battery pack, most periodic maintenance will be performed during scheduled
calibrations.

2.7 PARTICULATES—SAMPLED BY FILTRATION/IMPACTION
(FIGURE 2.37)

In sampling for particulates, the particulates must be filtered out or removed from the
airstream by impaction. Particulates that are suspended in the airstream come in many
sizes; therefore, the first question is whether exposure standards are based on the respirable
fraction or the total particulate levels.

Multiple-use calibration chambers may be used as protective environments around the
various filters, cyclones, and inhalable particulate monitors. These calibration chambers
effectively protect the particulate collection devices from extraneous particulate loading
during calibration cycles and airstream fluctuations. Total particulates are often analyzed
by gravimetric methods.

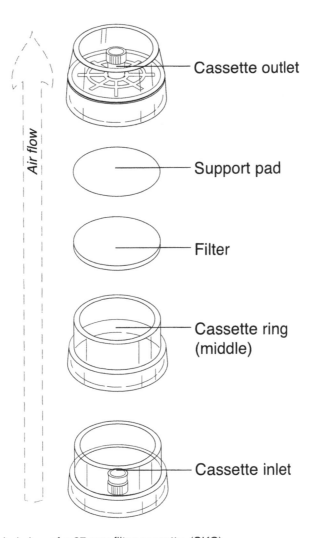

Figure 2.37   Exploded view of a 37-mm filter cassette. (SKC)

2.8 GRAVIMETRIC FILTER WEIGHING PROCEDURE

The step-by-step procedure for weighing filters depends on the make and model of the balance. Consult the manufacturer's instruction book for directions. In addition, follow these guidelines:

- Smoking and/or eating must not take place in the weighing area—both generate extraneous particulate matter in the airstream.
- Handle all filters with tongs or tweezers. Do not handle filters with bare hands.
- Desiccate all filters at least 24 hours before weighing and sampling. Change desiccant before the dessicant completely changes color (i.e., before the blue desiccant turns pink). Evacuate the desiccator with a sampling or vacuum pump.
- Zero the balance prior to use.
- Calibrate the balance prior to use and after every 10 samples.

- Immediately prior to placement on the balance, pass all filters over an ionization unit to remove static charges. (After 12 months of use, return the unit to the distributor for disposal.)
- Weigh all filters at least twice.
  —If there is more than a 0.005 mg difference in the two weighings, repeat the zero calibration and reweigh.
  —If there is less than a 0.005 mg difference in the two weighings, average the weights for the final weight.
- **Note:** To avoid damage to the weighing mechanism, take care not to exert downward pressure on the weighing pan(s).
- Record all the appropriate weighing information (in ink) in the weighing log.
- In reassembling the cassette assembly, remember to add the unweighed backup pad.

When weighing the filter after sampling, desiccate first and include any loose material from an overloaded filter and cassette.

2.9 TOTAL DUST AND METAL FUMES

A variety of filtration options is available to collect particulates. Sampling options are defined based on the regulatory onus and the sampling environment. Some examples of these options are as follows:

- Collect total dust on a preweighed, low-ash polyvinyl chloride (PVC) filter at a flow rate of about 2 l/min, depending on the rate required to prevent overloading. Weigh PVC filters before and after taking the sample.
- Collect metal fumes on a 0.8-μm mixed cellulose ester filter at approximately 1.5 l/min, not to exceed 2.0 l/min.
- When the gravimetric weight needs to be determined for welding fumes, collect these fumes on a low-ash PVC filter.

Take care to avoid overloading the filter, as revealed by any loose particulate in the filter cassette housing.

Personal sampling pumps must be calibrated before and after each day of sampling, using a bubble meter method (electronic or mechanical) or the precision rotameter method (which has been calibrated against a bubble meter).

2.10 RESPIRABLE DUST

Respirable dust is a component of particulates in the airstream that will deposit within the gaseous exchange areas of the lung (Figure 2.38). Respirable particles are just the right size to travel with inspired air into the alveoli of the lung. Once in the alveoli, these particles may be a simple irritant or may dissolve and, thus, become chemicals in suspension with tissue fluids. These suspended chemicals are then available to exert toxic and carcinogenic effects.

Respirable dusts that do not go into solution pose another danger. These insoluble dusts/particulates/fibers associated with respirable dusts are easy to breathe in, proceed with ease deeply into the lung, and once in the lung may stay in the tissue bed forever.

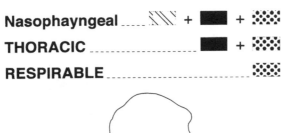

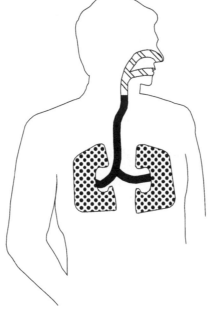

Figure 2.38 Inhalable particulate dust particles have a 50% cut point of 100 μg and are hazardous when deposited anywhere in the respiratory tract. Thoracic particulate particles have a 50% cut point of 10 μg and are hazardous when deposited anywhere in the lung airways and gas-exchange regions. Respirable particulate dust particles have a 50% cut point of 4 μg and are hazardous when deposited anywhere in the gas exchange regions. (SKC)

- Inert minerals such as asbestos cause fibrosis formation within the lungs by just mechanically irritating surrounding tissue.
- Other not so inert chemicals may produce biochemical effects as the body heats up the formerly semivolatile chemicals adsorbed or absorbed on the respired particulates. PAHs off-gas in the lung and become biochemically available through this body heat phenomenon.

For total particulate sampling results the "guess" is that 60% of the particles available in the airstream are ultimately respirable. The cut point for these particles is 50% at 4 μm. When health effects and exposure limits are based on respirable dusts,

- The 60% of total assumption must be made.
- Special instrumentation must be used to segregate out only the respirable fraction of total dust (see Figures 2.39, 2.40, and 2.41).

2.10.1 Cyclones

Cyclones of various types (aluminum, plastic) are used to collect respirable dust fractions. Plastic cyclones are the only choice in acid-gas contaminated atmospheres (Figures 2.42, 2.43, and 2.44). Collect respirable dust using a clean cyclone equipped (see

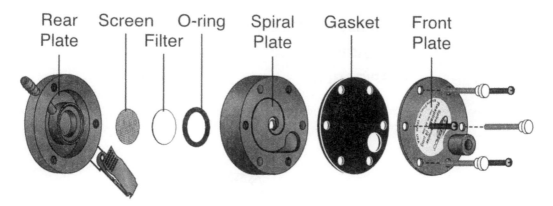

Figure 2.39 Exploded view of a Spiral Sampler. (SKC)

Figure 2.45) with a preweighed low-ash PVC filter (Figure 2.46). The flow rate should be 1.5–1.9 l/min.

2.10.1.1 Silica Respirable Dust—Cyclone Collection

Collect silica only as a respirable dust. Aluminum cyclones are recommended to ensure that the cyclone material does not interact or become part of the sample (Figures 2.47 and 2.48). Silica at sufficient velocity may etch a plastic cyclone.

A bulk sample should also be submitted to provide a basis for comparing silica levels in stock to ultimate respirable levels of dust. All filters used must be pre- and postweighed.

Calibration Procedures

1. Calibrate at the pressure and temperature where the sampling is to be conducted.
2. For respirable dust sampling using a cyclone, or for total dust sampling using an open-face filter cassette, set up the calibration apparatus as instructed.
3. Place the open-face filter cassette or cyclone assembly in a 1-l jar. The jar is provided with a special cover (Figure 2.49). If an aluminum cyclone is used, an aluminum cyclone calibration chamber can also be used in lieu of a 1-l sampling chamber (Figure 2.50).
4. Connect the tubing from the electronic bubble meter to the inlet of the jar.
5. Connect the tubing from the outlet of the cyclone holder assembly or from the filter cassette to the outlet of the jar and then to the sampling pump.
6. Calibrate the pump. Readings must be within 5% of each other.

The cyclone and filter cassette are now ready to be used. A holder makes placement of this assembly possible both for personnel and area sampling needs (Figure 2.51).

2.10.1.2 Cyclone Cleaning

For cyclone cleaning the following is required:

- Unscrew the grit pot from the cyclone.
- Empty the grit pot by turning it upside down and tapping it gently on a solid surface.
- Clean the cyclone thoroughly and gently after each use in warm soapy water or, preferably, wash in an ultrasonic bath.

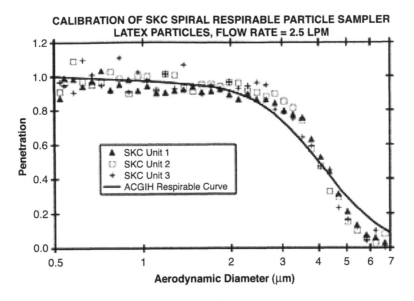

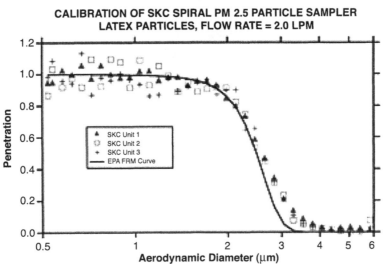

Figure 2.40 Calibration curves from Aerosol Dynamics, Inc. of the SKC spiral particle sampler PM2.5 using latex particles at 2.0 l/min. The data show a reasonable fit to the American Conference of Governmental Industrial Hygienists (ACGIH) respirable curve with a 4 μm cut point. (SKC)

- Rinse thoroughly in clean water, shake off excess water, and set aside to dry before reassembly.
- Never insert anything into the cyclone during cleaning.
- Inspect the cyclone parts for signs of wear or damage such as scoring, rifling, or a loose coupler.
- Replace units or parts if they appear damaged.
- Leak test the cyclone at least once a month with regular usage.

Figure 2.41   Spiral Sampler for respirable dust (a model available for PM2.5). (SKC)

Figure 2.42   The GS cyclone is a conductive plastic unit that holds a filter cassette for collecting respirable dust particles. The GS cyclone has a 50% cut point of 4.0 μm at 2.75 l/min. (SKC)

2.11 INHALABLE DUSTS (FIGURES 2.52 AND 2.53)

Inhalable dusts include all of those dusts from the general airstream that normal humans can bring into their respiratory tracts. The respiratory tract includes everything from the nose to the base of the lungs (Figure 2.54).

Inhalable dusts have a 50% cut point of 100 μm. Special inhalable dust samplers are used to collect only inhalable dusts; these samplers may vary according to the size of particulates collected (Figures 2.55, 2.56, and 2.57).

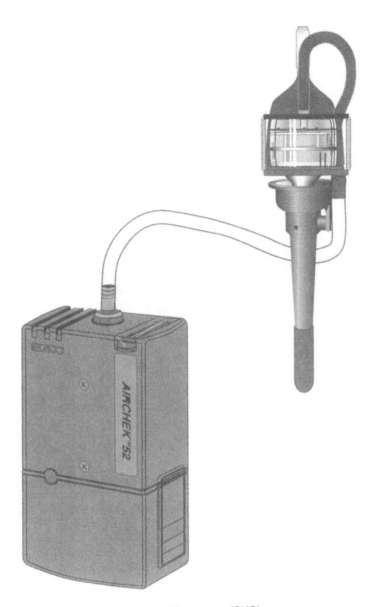

Figure 2.43   GS Cyclone attached to a sampling pump. (SKC)

2.12 PERSONNEL ENVIRONMENTAL MONITORS (PEMs)

For particulate segregation at either the 2.5 μm or 10 μm size, special personnel monitors are available. These PEMs use a single-stage impaction method to select particle size. The name is indicative of both personnel exposure concerns and the Environmental Protection Agency (EPA) particulates of concern given in the Clean Air Act requirements—thus personnel and environmental (Figures 2.58 and 2.59)!

Figure 2.44 The conductive plastic respirable dust cyclone is a lightweight conductive plastic unit. The unit is designed for a 50% cut point of 5.0 μm at 1.9 l/min and 4.0 μm at 2.2 l/min. Conductive plastic construction eliminates static problems. (SKC)

2.13 WELDING FUMES

When sampling for welding fumes, the filter cassette must be placed inside the welding helmet to obtain an accurate measurement of the employee's exposure. If, however, the welding helmet cannot be used as a sampling environment, the personal sampling pump cassette can be attached in the breathing zone at collar level. The resulting information can be used as a screening tool: "The air outside the helmet was (not) at a level of concern; therefore, the air inside the welding helmet was (not) at a level of concern."

Welding fume samples are normally taken using 37-mm filters and cassettes; however, if these cassettes will not fit inside the helmet, 25-mm filters and cassettes can be used. Care must be taken not to overload the 25-mm cassette when sampling.

2.14 ASBESTOS

Collect asbestos on a special 0.8-μm pore size, 25-mm diameter mixed cellulose ester filter with a backup pad.

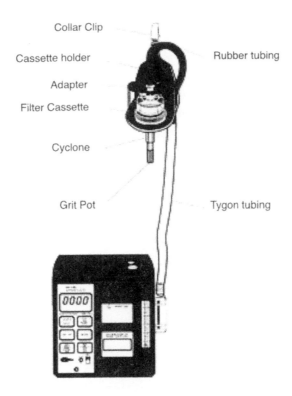

Figure 2.45 A cyclone attached to air sampling pump. (SKC)

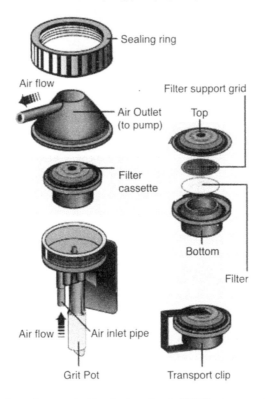

Figure 2.46 Exploded view of a respirable dust cyclone. (SKC)

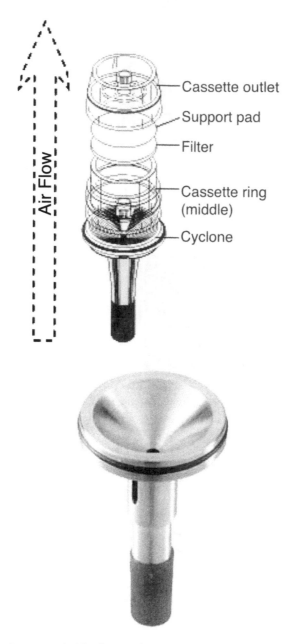

Figure 2.47 Two aluminum respirable dust cyclone models can be used with a 25- or 37-mm filter loaded onto a three-piece filter cassette. The cyclone separates the dust particles according to size. The respirable dust particles collect on a filter for analysis, while the larger dust particles fall into the grit pot and are discarded. (SKC)

- Use a fully conductive cassette with conductive extension cowl.
- Sample open face in worker's breathing zone.
- Ensure that the bottom joint (between the extension and the conical black piece) of the cassette is sealed tightly with a shrink band or electrical tape.
- Point the open face of the cassette down to minimize contamination.

Figure 2.48 Exploded view of a respirable dust cyclone cassette assembly. (SKC)

Figure 2.49 A calibration chamber allows calibration when using a cyclone, PUF tube, OVS tube, the IOM, or PEM. (SKC—Multipurpose Calibration Chamber)

- Use a flow rate of 0.5–5 l/min; 1 l/min is suggested for general sampling. Office environments allow up to 5 l/min.
- Calibrate pump before and after sampling. Calibration may be done with the cassette and cyclone replaced by the asbestos filter cassette (Figure 2.60).

Figure 2.50   Aluminum cyclone calibration chamber. (SKC)

Figure 2.51   Cassette holder on aluminum cyclone with filter cassette (37 mm). (SKC)

- Do not use nylon or stainless-steel adapters if in-line calibration is done.
- Sample for as long a time as possible without overloading (obscuring) the filter.
- Instruct the employee to avoid knocking the cassette and to avoid using a compressed-air source that might dislodge the sample while sampling.
- Submit 10% blanks, with a minimum in all cases of 2 blanks.

Where possible, collect and submit to the laboratory a bulk sample of the contaminant suspected to be in the air.

Figure 2.52   Inhalable dust sampler. Simulates dust collection of the nose and mouth. Meets NIOSH method 5700 sampling criteria for formaldehyde on dust. (SKC—IOM Inhalable Dust Sampler)

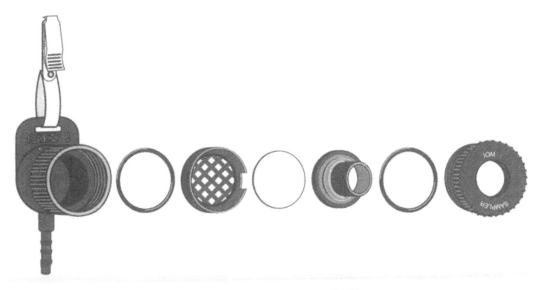

Figure 2.53   Exploded view of IOM inhalable dust sampler. (SKC)

About Dust Exposure

An individual's personal dust exposure does not depend solely on dust in the outside air. People now spend less time outdoors; therefore, indoor sources of dust particles can be just as important. In the future, personal measurement of PM2.5 may become more important than ambient air dust monitoring for PM2.5 for health effects studies. The EPA plans to examine PM2.5 using personal dust sampling devices in microenvironments—inside and outside schools, homes, and workplaces.

About Particulates

The 4 μm convention is based on respirability of all dust particles no matter the source. However, the PM2.5 convention separates dust particles based on likely sources. The most significant contributors to dust particles smaller than 2.5 μm include mechanical processes such as rock weathering that causes windblown dust, rock crushing, building demolition, and home remodeling (plaster and wood dust).

Figure 2.54   Information provided by SKC.

Figure 2.55   Sampling train using Button inhalable dust sampler. (SKC)

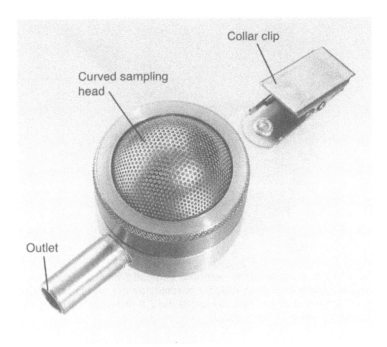

Figure 2.56 A Button aerosol inhalable dust sampler is a reusable filter sampler with a porous curved-surface sampling inlet designed to improve the collection characteristics of inhalable dust particles (<100 μm in aerodynamic diameter). (SKC)

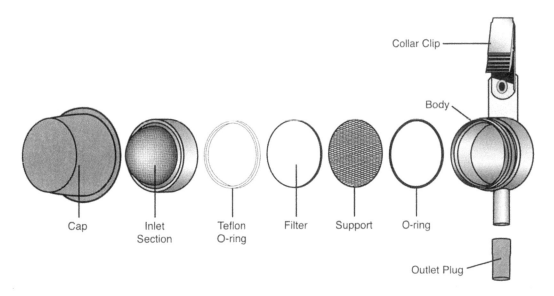

Figure 2.57 Exploded view of the Button aerosol inhalable dust sampler. (SKC)

Figure 2.58   The PEM is a lightweight, personal sampling device consisting of a single-stage impactor and a final filter. Aerosol particulates are sampled through the single-stage impactor to remove particulates above the 2.5 or 10 μm cut point, depending on which sampling head is chosen. (SKC)

2.15 DIRECT-READING DUST MONITORS (FIGURE 2.61)

2.15.1 Condensation Nuclei Counters (CNCs)

A CNC is a miniature, continuous-flow counter that takes particles too small to be easily detected, enlarges them to a detectable size, and counts them. Submicrometer particles are grown to supermicrometer alcohol droplets by first saturating the particles with alcohol vapor as they pass through a heated saturator lined with alcohol soaked felt and then condensing the alcohol on the particles in a cooled condenser. Optics focus laser light into the sensing volume.

As the droplets pass through the sensing volume, the particles scatter the light. The light is directed onto a photodiode that generates an electrical pulse from each droplet. The concentration of particles equals the number of pulses generated.

The counter counts individual airborne particles from sources such as smoke, dust, and exhaust fumes. It operates in three modes, each with a particular application. In the "count" mode the counter measures the concentration of these airborne particles. In the "test" (or fit test) mode measurements are taken inside and outside a respirator, and a fit factor is calculated. In the "sequential" mode the instrument measures the concentration on either side of a filter and calculates filter penetration.

This instrument is sensitive to particles as small as 0.02 μm, yet it is insensitive to variations in size, shape, composition, and refractive index.

An example of this type of monitor is the PortaCount used to determine particulate levels during quantitative fit testing of respirators. Because of shipping regulations for flammable liquids, reagent-grade isopropyl alcohol may have to be purchased locally and used to refill the small plastic alcohol-fill tubes provided with the PortaCount. CTC also stocks and ships this alcohol. For long-term storage (over 14 days), follow the steps listed below:

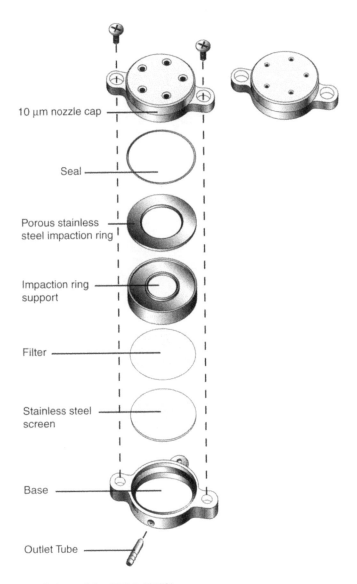

10 µm nozzle cap

Seal

Porous stainless
steel impaction ring

Impaction ring
support

Filter

Stainless steel
screen

Base

Outlet Tube

Figure 2.59 Exploded view of the PEM. (SKC)

- Dry the saturator felt after installing a freshly charged battery.
- Pack without adding alcohol.
- Allow the instrument to run until the *LO* message (low battery) or the e—e message (low particle count) appears.
- Remove the battery pack from the PortaCount.
- Install the tube plugs into the ends of the twin-tube assembly.

2.15.1.1 *Calibration*

Calibrate the counter before and after each use in accordance with the manufacturer's instructions.

Figure 2.60 A sampling pump can be used in high flow (750–5000 ml/min) or low flow (5–500 ml/min). Low flow requires an adjustable low-flow holder. (SKC—Model 224-PCXR8)

2.15.1.2 Maintenance

Isopropyl alcohol must be added to the unit every 5 to 6 hours of operation, per the manufacturer's instructions. Take care not to overfill the unit.

Under normal conditions a fully charged battery pack will last for about 5 hours of operation. Low-battery packs should be charged for at least 6 hours; battery packs should not be stored in a discharged condition.

2.15.1.3 Photodetection

Photodetectors operate on the principle of the detection of scattered electromagnetic radiation in the near infrared (Figures 2.62 and 2.63). Photodetectors can be used to monitor total and respirable particulates. The device measures the concentration of airborne particulates and aerosols, including dust, fumes, smoke, fog, mist. Certain instruments have been designed to satisfy the requirements for intrinsically safe operation in methane-air mixtures.

2.15.1.4 Calibration

Factory calibration is adequate.

2.15.1.5 Maintenance

When the photodetector is not being operated, it should be stored and sealed in its plastic bag to minimize particle contamination of the inner surfaces of the sensing chamber.

Figure 2.61 Real-time dust monitoring offers immediate real-time determinations and data recordings of airborne particle concentration in milligrams per cubic meter. Interchangeable size-selective sampling heads allow PM10, PM2.5, or PM1.0 monitoring. (HAZ-DUST—EPAM-5000 Environmental Particulate Air Monitor) (SKC)

After prolonged operation in or exposure to particulate-laden air, the interior walls and the two glass windows of the sensing chamber may become contaminated with particles. Repeated updating of the zero reference following the manufacturer's procedure will correct errors resulting from such particle accumulations. However, this contamination could affect the accuracy of the measurements as a result of excessive spurious scattering and significant attenuation to radiation passing through the glass windows of the sensing chamber.

2.15.2 Diesel Particulate Matter (DPM)

Sampling for particulates is sometimes accomplished using surrogate sampling and analysis. In DPM sampling elemental carbon is used as the basis for evaluating DPM levels (Figures 2.64 and 2.65).

Figure 2.62   Handheld dust monitors are available in handheld sizes. (SKC)

Figure 2.63   This real-time dust monitor can be worn as a personal monitor. (SKC)

Figure 2.64   NIOSH Method 5040 recommends sampling for DPM by elemental analysis of carbon as a surrogate. A cyclone is used with a 37-mm heat-treated quartz filter and cellulose support pad that has been prepared in a carbon-free humidity-controlled clean room. (SKC)

Figure 2.65   DPM sampler assembly in cassette holder. (SKC)

2.16 BIOLOGICALS

The world of biological risk assessment is a new and challenging one. Until recently we did not have dynamic methods to test many areas for biological risk. Essentially we relied on shipping sampled material and swabs to laboratory sites for culturing. In some cases we were able to obtain air samples using filter cassettes or impingement onto agar plates with an Anderson air-sampling tower.

All of these methods are still used; however, a new tool is now available—an air-sampling device developed to meet the needs of clean rooms and immune-suppressed patient care medical facilities. With this new sampling device, the Reuter Centrifugal Sampler (RCS) system, we can insert an instrument the size of a handheld vacuum into locations formerly inaccessible. We can obtain a sample with a designated flow of air, e.g., 50–1000 m^3, as a further quantification aid for samples. The samples are impinged upon air-sampling agar plates, where growth may immediately begin. Transfer to laboratory sites allows controlled further growth in biologically safe cabinets.

Contact and liquid dip agar plates are used to compare these results to airborne levels. These plates are secured for future laboratory incubation and analysis. At no time are these or the air-sampling media plates left unattended prior to transfer to the laboratory, thereby keeping chain-of-custody intact.

Because, unlike chemicals, biologicals can and do multiply through various reproduction means, the use of personnel protective equipment (PPE) is always a requirement during sampling.

2.16.1 General Sampling Protocols

The following are step-by-step procedures for mold sampling:

1. Assemble materials and equipment to be used. Segregate materials and equipment to be taken inside the building or area of concern. Use impermeable plastic bags whenever possible to containerize materials and equipment to be taken into the building. Do not use cardboard or other porous containers that cannot be readily decontaminated.
2. Mark each contact sample or strip agar blister pack with a unique sample number using a Sharpie pen. Allow the ink to air dry before overpackaging the blister pack in a baggie.
3. Using a quart or larger size Ziploc freezer bag; overpackage each contact sample or strip agar.
4. Assemble at least 10 of each type of sampling media (contact strips for the RCS) in a large overpackaged baggie. Package no more than 10 agar blister packs together for transfer to a contaminated area.
5. Assemble another bag to contain extra impermeable gloves (latex 6 mil or neoprene) and alcohol wipes. Alcohol wipes can be purchased in small manufactured packages or made up on-site using paper towels and isopropyl alcohol. The made up on-site alcohol towels are more effective for larger decontamination areas. Double bag all sources of alcohol and avoid direct alcohol contact with the agar blister packs.
6. Establish a staging area and set up a decontamination area in a biologically neutral location away from potential biological amplification sites.
7. Don PPE in the following order:
   - Don first hooded Tyvek.
   - Don boots.
   - Don first and second layer of gloves (double gloving is optional in some situations).
   - Duct tape boot/glove openings at ankles/wrists (optional in some situations).
   - Don respirator.
   - Don second hooded Tyvek (optional in some situations).
8. Begin the sampling routine. Sample outside and in all assumed uncontaminated or amplified areas first; then sample into progressively more contaminated areas.

Use the same protocols for all sampling events, including the same pressure and motion when using contact plates and the same walking routines or static placement when using the RCS.

2.16.2 Contact and Grab Sampling

Contact and specimen grab-sampling routines are as follows:

1. Open sample overbag at first location to be sampled.
2. Sample mold by applying the contact plate to the area with some pressure, by swabbing, or by obtaining a small sample of contaminated building (or other) material.
3. Place mold-contaminated item into the sample bag or swab vial.
4. Seal the overbag.
5. Decontaminate gloves if contaminated by direct contact or if pathogenic fungi are suspected.
6. Place used decon pad into small waste bag.
7. Decontaminate spatula or any other tools used.
8. Place decon pad (if used) into small waste bag. Decon pads can be baggies or small pieces of precut plastic.
9. Decontaminate outside of sealed sample bag if contaminated by direct contact or if pathogenic fungi are suspected.
10. Place used decon pad into small waste bag.
11. Repeat steps 1–10 for additional sampling locations.

2.16.3 Reuter Central Fugal System (RCS)

Both high-volume and low-volume RCS units are available (see Figures 2.66 and 2.67). The RCS units can be mounted on stands if necessary (Figure 2.68). RCS sampling routines are as follows:

Figure 2.66   Biological air sampling instrument. (Biotest Diagnostic Corp. Air Sampler RCS)

Figure 2.67   RCS centrifugal air sampler. (Biotest Diagnostic Corp.)

Figure 2.68   Tripod and remote can be used to sample for biological contamination in ventilation systems. (Biotest Diagnostic Corp.)

1. Open sample overbag at first location to be sampled.
2. Insert RCS agar strip into the RCS impeller assembly. Do not directly touch the agar medium at any time. In the event that the agar is touched, discard that agar strip.
3. Sample mold by running the RCS for the approved time duration.
4. Remove the RCS agar strip from the RCS impeller assembly. Do not directly touch the agar medium at any time. In the event that the agar is touched, discard that agar strip.
5. Place the RCS agar strip into the original sample overbag.
6. Seal the overbag.
7. Decontaminate gloves if contaminated by direct contact or if pathogenic fungi are suspected.
8. Place used decon pad into small waste bag.
9. Decontaminate outside of sealed sample bag if contaminated by direct contact or if pathogenic fungi are suspected.
10. Place used decon pad into small waste bag.
11. Repeat steps 1–10 for additional sampling locations.

The RCS in some circumstances may need to be decontaminated between sampling events. In the field the impeller assembly can be cleaned with isopropyl alcohol and thoroughly air-dried in a biologically neutral location. If further decontamination is required, the RCS will need to be decontaminated at the issuing laboratory.

In some circumstances pathogenic sleeves must be used with the RCS. Do not take the RCS carrying case or battery charger into a contaminated environment.

At the conclusion of a sampling event, at a minimum, wipe down the RCS exterior with alcohol wipes. (Contact a CIH for additional decontamination requirements.)

Use equipment decon pads to decontaminate temporary lighting and any other large equipment used. **Note:** Lights are turned off prior to decontaminating. The last set of lights may be decontaminated using handheld flashlights for illumination.

2.16.4 Exit Requirements

When exiting the area,

- Seal all used interior decon pads in small waste bag.
- Exit area.
- Decontaminate outer Tyvek and respirator with decon wipes.
- Remove outer Tyvek.
- Place used Tyvek into large waste bag.
- Decontaminate inner Tyvek, gloves, and boots.
- Place used decon pads into waste bag.
- Remove duct tape from wrists/ankles.
- Remove boots, gloves, and Tyvek.
- Place used boots, gloves, and Tyvek into large waste bag.
- Seal large waste bag.
- Decontaminate respirator again prior to removal. Place used decon pads into (new) small waste bag.
- Remove respirator.
- Use decon pads to decontaminate hands.
- Place all used decon pads into small waste bag.
- Bag all disposable equipment for disposal in an approved manner (contact CIH for project specific advice as to disposal).
- Bag all nondisposable equipment for further decontamination off-site (contact CIH for project specific advice as to disposal).

2.16.5 Static Placement Impingement

Less mobile sampling devices are also available. These include the use of filter cassettes to collect all spores—viable and nonviable. The Anderson sampling device has various impaction trays installed to segregate spores on the basis of size.

2.16.6 Bioaerosols

When bioaerosols must be collected, extremely high flow rates may be required. The rule in general is that sonic flow requires a 0.5 atm pressure drop (Figure 2.69). As with all pumps, the greater the pressure drop, the faster the intake of air toward that pressure void area.

Figure 2.69 A noncompensating vacuum pump is capable of maintaining the 0.5 atm pressure drop required for sonic-flow applications. (SKC—Vac-U-Go Sampler)

2.17 RADIATION MONITORS AND METERS

2.17.1 Light Meter

The light meter is a portable unit designed to measure visible, UV, and near-UV light in the workplace.

The light meter is capable of reading any optical unit of energy or power level if the appropriate detector has been calibrated with the meter. The spectral range of the instrument is limited only by the choice of detector.

Steady-state measurements can be made from a steady-state source using the "normal operation" mode. Average measurements can be read from a flickering or modulated light source with the meter set in the "fast function" position. Flash measurements can be measured using the "integrate" function.

Exposure of the photomultiplier to bright illumination when the power is applied can damage the sensitive cathode or conduct excessive current.

2.17.1.1 Calibration

No field calibration is available. These instruments are generally very stable and require only periodic calibration at a laboratory.

2.17.1.2 Maintenance

Little maintenance is required unless the unit is subjected to extreme conditions of corrosion or temperature. Clean the optical unit with lens paper.

Detector heads should be recalibrated annually by the manufacturer only. All calibrations are National Institute of Standards and Technology (NIST) traceable.

The Ni-Cad batteries can be recharged. Avoid overcharging, which will reduce battery life.

2.18 IONIZING RADIATION

Because ionizing radiation cannot be detected by the human senses, detection and quantification must be accomplished by specifically designed instruments. All such methods of measurement employ a substance that responds to radiation in a measurable way and a system or apparatus to measure the extent of the response. Most radiation detectors operate by one of two methods: ionization or scintillation.

The selection of instruments is based on the type and energy range of the radiation expected on-site. Survey instruments will be chosen for their sensitivity to the type of radiation present in the area to be surveyed. The method of relating the instrument reading to milliroentgens per hour will be included.

Some instruments can measure multiple types of radiation and require methods for determining which type of radiation is being measured. For example, neutron detectors for neutron dose can be quantified using boron trifluoride (BF_3) detectors.

The ionizing radiation survey meter is useful for measuring radon decay products from air samples collected on filters.

- The barometric pressure should be noted for ionizing radiation chambers.
- Wipe samples collected on a filter can also be counted with this detector, and general area sampling can be done.
- The survey meter with the scintillation detector can be used to measure the presence of radon-decay products in a dust sample.

2.18.1 Ionization Detectors

Most ionization detectors consist of a gas-filled chamber with a voltage applied; a central wire becomes the anode, and the chamber wall becomes the cathode. Any ion pairs produced by radiation interacting with the chamber move to the electrodes, where they are collected to form an electronic pulse that can be measured and quantified. Depending on the voltage applied to the chamber, the detector may be considered a gas proportional detector, a Geiger-Muller (GM) detector, or an ion chamber.

2.18.1.1 Gas Proportional Detectors

Thin-window gas proportional detectors may be used to detect alpha and beta radiation. Distinction between alpha and beta is achieved by adjusting the voltage of the detector.

2.18.1.2 Ion Chamber

An ionization chamber is a gas-filled chamber containing an anode and a cathode. As radiation passes through the gas, it ionizes some of the gas molecules. These ion pairs are attracted to the anode and cathode and create an electrical pulse. The pulses are counted, integrated, and displayed on the meter face in roentgens per hour. Ion chambers provide a nearly linear response to gamma and X-ray radiation above a few kiloelectron volts in energy and at radiation levels above 0.1 mR/h. For this reason an ion chamber is the only instrument for quantifying radiation exposures. Ion chambers may be used to quantify the beta, gamma, or X-ray dose at a location.

Sodium iodide (NaI) scintillation detectors, used on survey meters, provide better counting efficiency for gamma and X-rays, but have a more limited range of energies, depending on the size of the crystal and the density of the window. NaI detectors may be used to detect the presence of gamma radiation, but only if the energy level of the radiation is known, and the correct size crystal is used.

2.18.1.3 GM Detector

Because of its versatility and dependability, the GM detector is the most widely used portable survey instrument. A GM detector with a thin window can detect alpha, beta, and gamma radiation. The GM is particularly sensitive to medium-to-high energy beta particles (e.g., as from ^{32}P), X-rays, and gamma rays. The GM detector is fairly insensitive to low-energy X-rays or gamma rays, such as those emitted from ^{125}I, and to low-energy beta particles, such as those emitted by ^{35}S and ^{14}C; it cannot detect the weak betas from ^{3}H.

Unlike the ion chamber the GM detector does not actually "measure" exposure rate. It instead "detects" the number of particles interacting in its sensitive volume per unit time. The readout of the GM is in counts per minute (cpm), although it can be calibrated to approximate milliroentgens per hour for certain situations. With these advantages and limitations a GM detector on a rugged survey meter is the instrument of choice for initial entry and survey of field radiation sources and radioactive contamination.

GM detectors are calibrated to one energy level of the electromagnetic spectrum—usually 662 keV, the gamma energy from the decay of ^{137}Cs, and are read out in milliroentgens per hour. GM detector efficiency for radiation at other energies is not linear. GM detectors can be used to detect the presence of radiation, but only as a rough estimate of the dose rate. Beta shields will compensate for blocking betas and reading gamma or X-ray radiation. Subtraction of the gamma or X-ray radiation readings will yield an approximation of the beta contribution. Beta contribution should be read in counts per minute.

2.18.2 Scintillation Detectors

Scintillation detectors are based upon the use of various phosphors (or scintillators) that emit light in proportion to the quantity and energy of the radiation they absorb. The light flashes are converted to photoelectrons that are multiplied in a series of dynodes (i.e., a photomultiplier) to produce a large electrical pulse. Because the light output and resultant electrical pulse from a scintillator is proportionate to the amount of energy deposited by the radiation, scintillators are useful in identifying the amount of specific radionuclides present (i.e., scintillation spectrometry).

Solid scintillation detectors are particularly useful in identifying and quantifying gamma-emitting radionuclides. NaI scintillation detectors, used on survey meters, provide better counting efficiency for gamma and X-rays, but have a more limited range of energies, depending on the size of the crystal and the density of the window. NaI detectors may be used to detect the presence of gamma radiation, but only if the energy level of the radiation is known, and the correct size crystal is used.

The common gamma counter employs a large (e.g., 2″ × 2″ or 3″ × 3″) NaI crystal within a lead-shielded well. The sample vial is lowered directly into a hollow chamber within the crystal for counting. Such systems are extremely sensitive, but do not have the resolution of more recently developed semiconductor counting systems. Portable scintillation detectors are also widely used for conducting various types of radiation surveys. Of particular use to researchers working with radioiodine is the NaI thin crystal detector capable of detecting the emissions from ^{125}I with efficiencies nearing 20% (a GM detector is less than 1% efficient for ^{125}I).

The most common means of quantifying the presence of beta-emitting radionuclides is through the use of liquid scintillation counting. In these systems the sample and phosphor are combined in a solvent within the counting vial. The vial is then lowered into a well between two photomultiplier tubes for counting. Liquid scintillation counting has been an essential tool of research involving radiotracers such as ^{3}H, ^{14}C, ^{35}S, and [45]Ca.

One problem that occurs with liquid scintillation counting is determining the efficiency of the system. The low-energy photons produced by the beta particles interacting with the scintillation cocktail are easily shielded from the photomultiplier tubes due to optical and chemical quenching. To account for this artifact, a quench curve must be computed using a set of increasingly quenched standards and a method of determining a quench-indicating parameter for each standard. The quench curve is a graph for a certain nuclide of each standard's quench-indicating parameter (QIP) vs. the efficiency of the counter. The efficiency for an unknown sample is then determined by measuring its QIP, with the graph determining the counter's efficiency for that sample. Then use that quantity and the counts per minute of the sample to determine the disintegration rate (dpm) of the sample.

Fortunately, most modern counters compute and store quench curves from a single counting of a standard set, use an external standard to determine the QIP, and automatically output sample counts in disintegrations per minute.

The other common problem with liquid scintillation counting is chemiluminescence. Certain chemicals when mixed with some scintillation cocktails will cause the cocktail to emit photons, resulting in a higher count rate than actually exists from the radionuclide itself. **Rule of thumb:** Let scintillation vials wait for a few hours after mixing the cocktail and the material to be assayed to allow for the initial chemiluminescence to be exhausted.

2.18.3 Counting Efficiency

The purpose of radiation counting systems is to determine sample activity (microcuries or disintegrations per minute). However, because of numerous factors related to both the counting system and the specific radionuclide(s) in the sample, the radiation detector can never see 100% of the disintegrations occurring in the sample. The counts per minute displayed by the counter must therefore be distinguished from the disintegration rate of the sample.

The ratio of the count rate to the disintegration rate expressed as a percent is the efficiency of the counting system.

$$\text{cpm/dpm} \times 100\% = \text{efficiency} \qquad (1)$$

Because every counting system will register a certain number of counts from environmental radiation and electronic noise in the counter (i.e., background), a more correct formula is as follows:

$$[(\text{Sample cpm} - \text{bkg cpm})/\text{Sample dpm}] \times 100\% = \text{efficiency} \qquad (2)$$

Example 1

A 1-ml sample of a solution of ^{125}I is counted in a gamma scintillation spectrometer with a window of 15 keV to 75 keV. A 0.1 μCi ^{125}I standard and a background sample are counted along with the solution for 1-min each. The results are as follows:

$$0.1 \ \mu Ci = 2,220,000 \ dpm = 2.22 \times 10^6 \ dpm/\mu Ci$$

$$Background \ (bkg) = 35 \ cpm$$

$$Standard = 110,658 \ cpm$$

$$Sample = 3,246,770 \ cpm$$

Efficiency of the counter:

$$= [(Sample \ cpm - bkg \ cpm)/Sample \ dpm] \times 100\% = efficiency$$

$$= [(110,658 - 35)/(0.1 \times 2.22 \times 10^6 \ dpm/\mu Ci)] \times 100\% = 0.4983$$

$$= 49.8\% \ efficiency \ of \ the \ counter$$

Activity of the sample:

$$= (Sample \ cpm - Background)/(efficiency \times dpm \ per \ \mu Ci)$$

$$= (3,246,770 - 35)/[0.4983 \times (2.22 \times 10^6 \ dpm/\mu Ci)]$$

$$= 2.935 \ \mu Ci/ml \ of \ sample$$

Example 2

A sample containing a ^{14}C-labeled amino acid is counted in a liquid scintillation counter. The sample count rate is 1200 cpm, and the background is 30 cpm. If the counter is 85% efficient for ^{14}C, what is the activity within the sample?

$$Sample \ cpm - bkg \ cpm/Sample \ dpm = efficiency$$

$$= (1200 - 30)/85 = 1376$$

$$1376 \ dpm \times 2.22 \times 10^6 \ dpm/\mu Ci$$

$$= 6.2 \times 10^{-4} \ \mu Ci$$

2.18.4 Monitoring for Radioactive Contamination

Monitoring instruments must be chosen for their sensitivity to the type of radiation to be monitored. A method of relating the instrument reading to microcuries must be included.

Monitoring must be performed slowly and at distances of 1 to 2 mm from the surface to detect low-intensity radiations. The monitoring results must be documented and must include the following:

- Identification of the instrument used to monitor
- Name of the person performing the monitoring
- Location, date, and time of monitoring
- Results of the monitoring
- Comments on any factors that might influence the readings

2.18.5 Daily Use Checks

Each survey instrument must have an appropriate check source attached or assigned to the instrument. The check source for the instrument must be surveyed immediately after calibration, and the reading must be written on the calibration sticker on the instrument.

Before each use of the instrument, the check source must be monitored, and the reading compared to the reading noted on the calibration sticker. Any meter not measuring within ±10% of the reading on the calibration sticker must be tagged as requiring maintenance and must not be used until maintenance and recalibration have been performed. Surveys must be performed slowly; the instrument needs time to integrate and display the measurement.

Survey results must be documented and must include the following:

- Identification of the instrument used to perform the survey
- Name of the person performing the survey
- Location, date, and time of the survey
- Results of the survey
- Comments on any factors that might influence the survey

2.18.6 Survey Instrument Calibration

Survey instruments must be calibrated periodically using procedures outlined in the instrument manual, within the following guidelines:

- All sources used to calibrate instruments must be traceable to NIST.
- The survey meter and the probe must be calibrated as a unit. If probes are changed, the unit must be recalibrated prior to use.
- Instruments must be calibrated at least annually.
- Instruments must be calibrated after every maintenance or repair operation.

The date of calibration, the date the next calibration is due, and the initials of the person performing the calibration must be written on a calibration sticker attached to the instrument.

2.19 NONIONIZING RADIATION

Electromagnetic fields (EMFs) produced by computer terminals, cellular phones, electric blankets, and power lines can be measured. Microwave radiation, whether used for communication, electron wave propulsion, or food warming, can also be measured.

EMFs are generated any time charged particles move through a medium. Two fields are actually produced, one electric and one magnetic. Both are always found together.

The earth itself is the largest source for a magnetic field, and lightning generates one of the strongest electric fields. Electricity transmitted through a wire produces an EMF proportional in size to the current flow and the voltage drop.

Another issue associated with EMFs is the corona phenomenon, where air around a high voltage power line is ionized and interferes with TV and radio reception. This ionization can also generate ozone that may be a health concern.

The electromagnetic spectrum is very broad and consists of low-frequency, low-energy waves, such as those generated by power lines, through high-energy, high-frequency cosmic rays.

- The wavelength of most interest is around 60 Hz. This frequency is the one most commonly used in the United States for electrical power transmission and falls into a range known as extremely low frequency EMF. The extremely low frequency EMF range is from 5 Hz to 2000 Hz. These waves are extremely long and can be easily shielded or attenuated.
- Cellular phones broadcast radio signals at around 850 MHz. These radio signals are equivalent to most television transmission frequencies. Cellular phones operate at a much higher frequency than most household sources of EMF.
- Ionizing radiation, such as X-rays, have a frequency of around 2.5×10^{15} Hz. Ionizing radiation, in very high doses, is known to lead to an increase in cancer incidence.

2.19.1 Guidance

No legally enforceable exposure standards are in place for any nonionizing radiation. The non-Ionizing Radiation Committee of the International Radiation Protection Association recommends the following standards for 60 Hz EMF:

Magnetic field strength:
- 5 Gs—occupational exposure
- 1 Gs—public exposure
Electric field strength:
- 10,000 V/m^3—occupational exposure
- 5,000 V/m^3—public exposure

Surveys of ten commonly used video display terminals gave readings in the following ranges:

- Magnetic field strength of 0.0006 to 0.0077 Gs
- Electric field strength of 1.6 to 15 V/m^3

2.19.2 Broadband Field Strength Meters

Broadband field strength meters are available for measuring electromagnetic radiation in the frequency range from 0.5 MHz to 6000 MHz. Each meter comes with probes for measuring either magnetic or electric field strength, batteries, headset, and carrying case.

This unit is designed for laboratory and field use to measure magnetic and electric fields near radio frequency (RF) induction heaters, RF heat sealers, radio and TV antennas, or any other RF sources.

- All units have automatic zeroing. There is no need to place the unit in a zero-field condition to zero it.
- All units have a peak memory-hold circuit that retains the highest reading in memory.
- All units operate with either electric (E) or magnetic (H) field probes based on diode-dipole antenna design. Total field strength is measured at the meter regardless of the field orientation or probe-receiving angle. The diode-dipole antenna design of the probe is much more resistant to burnout from overload than the thermocouple design of probes used with other meters.

2.19.2.1 Calibration

No field calibration is available. Periodic calibration by a laboratory is essential.

2.19.2.2 Maintenance

No field maintenance is required other than battery-pack charging or replacement.

CHAPTER **3**

Calibration Techniques

This chapter contains theoretical and real-world discussions about the intricacies of calibration. It gives special emphasis to situations where the knowledge of calibration techniques is a prerequisite for sampling adequacy.

Calibration is the means used to provide evidence that instrumentation is working accurately and that results are reliable and repeatable within the tolerance levels prescribed for these instruments. The calibration of real-time instruments is very specific, and manufacturer's instructions must be followed. Increasingly, instrumentation is calibrated using electronic circuitry, with empirical tests against standards in accordance with the manufacturer's requirements at prescribed intervals.

Collection efficiency must also be taken into account when calibration curves are developed. As filters, cartridges, or sorbent tubes load, air intake and calibrated airflow volumes across the media can be expected to change (Figure 3.1).

3.1 CALIBRATION REQUIREMENTS

Instrument calibration records must be reviewed periodically by the users to assess accuracy of documentation and to evaluate instrument performance. Assessment is based on the instrument operating instructions and knowledge of the user.

- Users will check minimum calibration frequency requirements for the instrument and calibrate according to the applicable operating instruction manual.
- Calibrating the instrument at the worksite under actual field conditions is an important requirement.
- Air-sampling instrument readings for calibration will be corrected for temperature and barometric pressure.

Record calibration and challenge results as follows in the field logbook and include the following information:

- Instrument identification
- Date (if not on page in logbook)
- Precalibration readings as found or as set
- Readings after calibration and span settings, if applicable

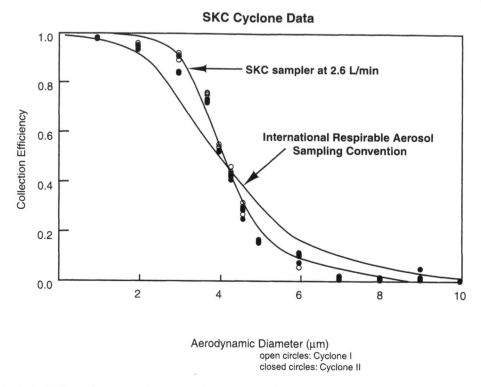

Figure 3.1   SKC cyclone sampler at 2.6 l/min collection efficiency vs. aerodynamic diameter.

- Calibration gas (challenge) used and lot number
- Signature of person performing calibration

3.1.1 Calibration Assurance

The user will transport the instrument to the field in its carrying case or by other means that will adequately protect the instrument. All devices to be used in the field sampling train must be in-line during calibration (Figure 3.2).

Relevant information such as the type of instrument to be used, frequency of monitoring, and specific warning and action levels are found in the site safety and health plan or Activity Hazard Analysis (AHA) written for the specific job site. The user must follow these requirements:

- Users are expected to follow the specific manufacturer's operating instruction manual for the instrument being used. The user takes a field copy of the operating instruction manual into the field with the instrument.
- Prior to entering a contaminated zone, the user will take appropriate precautions to prevent contamination of the instrument, e.g., using plastic bags or filters.
- The instrument is used in accordance with the applicable operations and maintenance (O&M) manual and site safety and health plan.
- Data are recorded in the field logbook.

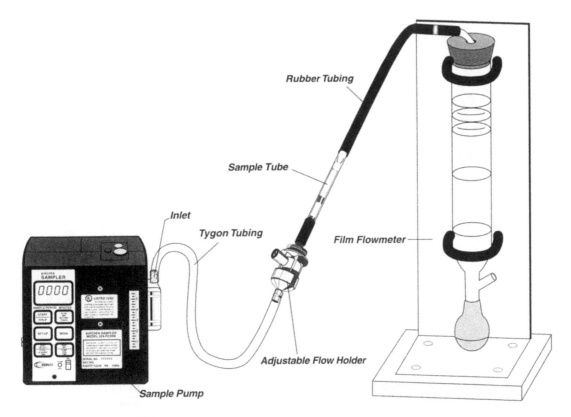

Figure 3.2   A primary standard flowmeter connected to a sampling train. (SKC)

3.1.2 Decontamination

Decontamination of instruments is necessary to remove contaminants present at a site. The user is expected to take the proper steps to assure the instrument is clean prior to leaving a contaminated zone. A radiation release sticker may be required depending on the circumstances of radioactive contamination on-site.

Decontamination procedures will vary depending on the individual contaminants and circumstances involved. Steps may be as simple as wiping down the instrument probe with soap and water or as involved as disassembly and thoroughly cleaning with decontamination agents.

Decontaminate the instrument in accordance with the site safety and health plan decontamination procedures or instructions provided by air monitoring professionals.

3.1.3 Maintenance

Users should not attempt field repairs other than preventive maintenance, cleaning, adjustments, or decontamination. Major repairs must be done by a qualified instrument technician or the factory. Refer to the O & M manual for maintenance instructions. Flow-rate holders also require frequent maintenance to assure consistent operation (Figure 3.3).

- *Battery charging* will be done in accordance with the manufacturer's O&M manual to prevent overcharging or possible damage to the equipment.

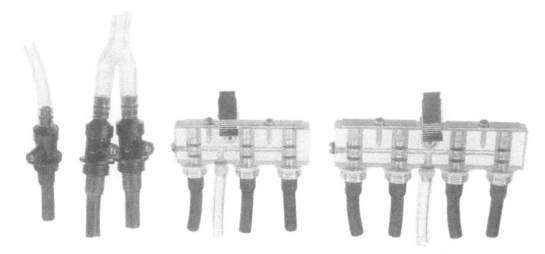

Figure 3.3   Single, double, triple, and quadruple adjustable low flow holders. (SKC)

- Manufacturer's preventive maintenance procedures and major repairs by the factory or instrument technician will be recorded in the field logbook or calibration logs for that instrument.
- After field maintenance, instrument calibration is invalidated, and recalibration is necessary prior to use.

The terms *primary* and *secondary* calibration devices are used in field calibration. Primary calibration means that a physical phenomenon whose progression is dictated by the laws of physics is measured for quantification. Such a physical phenomenon is the time required for a bubble to move up a glass tube or the time required for a gas to evaluate a chamber.

Secondary calibration involves the use of equipment previously calibrated against a primary standard. Such equipment is often the proper choice because primary standards often cannot be readily transported and relied upon under field conditions.

3.2 MANUAL BURET BUBBLE METER TECHNIQUE (PRIMARY CALIBRATION)

When a sampling train requires an unusual combination of sampling media (e.g., glass fiber filter preceding impinger), the same media/devices should be in-line during calibration. Calibrate personal sampling pumps before and after each day of sampling (Figure 3.4).

3.2.1 Bubble Meter Method

For the Bubble Meter Method use the following procedures:

1. Allow the pump to run 5 min prior to voltage check and calibration.
2. Assemble the polystyrene cassette filter holder using the appropriate filter for the sampling method.
3. If a cassette adapter is used, care should be taken to prevent contact with the backup pad.
4. **Note:** When calibrating with a bubble meter, the use of cassette adapters can cause moderate to severe pressure drop in the sampling train, which will affect the calibration result. If adapters are used for sampling, then they should be used during calibration.

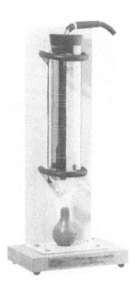

Figure 3.4 This portable standard flowmeter can be used in the office or laboratory, and in the field. (SKC)

5. Connect the collection device, tubing, pump, and calibration apparatus.
6. A visual inspection should be made of all Tygon® tubing connections.
7. Wet the inside of a 1-l burette with a soap solution.
8. Turn on the pump and adjust the pump rotameter to the appropriate flow-rate setting.
9. Momentarily submerge the opening of the burette in order to capture a film of soap.
10. Draw two or three bubbles up the burette so the bubbles will complete their run.
11. Visually capture a single bubble and time the bubble from 0 to 1000 ml for high-flow pumps or 0 to 100 ml for low-flow pumps.
12. The timing accuracy must be within +1 s of the time corresponding to the desired flow rate.
13. If the time is not within the range of accuracy, adjust the flow rate and repeat steps 9 and 10 until the correct flow rate is achieved. Perform steps 9 and 10 at least twice in any event.
14. While the pump is still running, mark the pump or record (on the calibration record form) the position of the center of the float in the pump rotameter as a reference.

Repeat the procedures described above for all pumps to be used for sampling. The same cassette and filter may be used for all calibrations involving the same sampling method.

Calibration of multiple tubes, whether in series or in parallel, is sometimes required. Any flow adjustment mechanisms and critical orifices must also be installed in the sampling train during these calibrations. When flexible tubing is used, it must be consistently used in all applications during sampling events (Figure 3.5).

3.3 ELECTRONIC FLOW CALIBRATORS

Electronic flow calibrators are high-accuracy, electronic bubble flowerets that provide instantaneous airflow readings and a cumulative averaging of multiple samples.

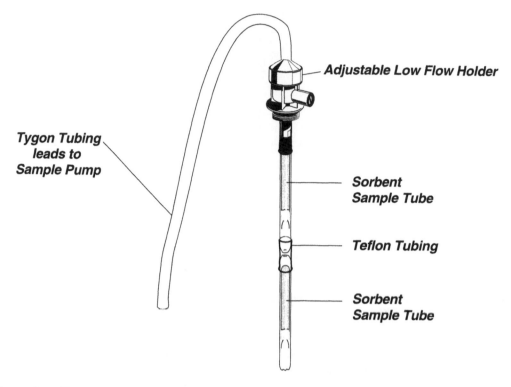

Figure 3.5   Two sorbent sampling tubes in series with a single adjustable low-flow holder. (SKC)

- These calibrators measure the flow rate of gases and present the results as volume per unit of time.
- These calibrators may be used to calibrate all air-sampling pumps.
- The timer is capable of detecting a soap film at 80-μs intervals.
- This speed allows under steady flow conditions an accuracy of ±0.5% of any display reading.
- Repeatability is ±0.5% of any display.
- The range with different cells is from 1 ml/min to 30 l/min.
- Battery power will last 8 h with continuous use. Charge for 16 h. These meters can be operated from A/C chargers.

When a sampling train requires an unusual combination of sampling media (e.g., glass fiber filter preceding impinger), the same media and devices should be in-line during calibration (Figure 3.6).

3.3.1 Cleaning Before Use

Before using an Electronic Flow Calibrator:

- Remove the flow cell and gently flush it with tap water. The acrylic flow cell can be easily scratched.
- Wipe with cloth only.
- Protect center tube, where sensors detect soap film, from dirt and scratches.
- *Never* clean with *acetone*. Use only soap and warm water.

Figure 3.6   The electronic flow calibrator using soap bubbles is easy to use and suitable for field use. (SKC)

When cleaning a flow cell prior to storage, allow it to air-dry. If stubborn residue persists, it is possible to remove the bottom plate. Squirt a few drops of soap into the slot between base and flow cell to ease residue removal.

3.3.2 Leak Testing

The system should be leak checked at 6-in. H_2O by connecting a manometer to the outlet boss and evacuate the inlet to 6 in. H_2O. No leakage should be observed.

3.3.3 Verification of Calibration

The calibrator is factory calibrated using a standard traceable to the NIST, formerly called the NBS.

- Attempts to verify the calibrator against a glass 1-l burette should be conducted at 1000 ml/min for maximum accuracy.
- The calibrator is linear throughout the entire range.

3.3.4 Shipping and Handling

When transporting, especially by air, one side of the seal tube that connects the inlet and outlet boss must be removed for equalizing internal pressure within the calibrator. Do not transport the unit with soap solution or storage tubing in place.

3.3.5 Precautions and Warnings

Avoid the use of chemical solvents on the flow cell, the calibrator case, and the faceplate. Generally, soap and water will remove any dirt.

- Never pressurize the flow cell at any time with more than 25-in. water pressure.
- Do not charge batteries longer than 16 h.
- Do not leave A/C adapter plugged into calibrator when not in use to avoid damage to the battery supply.
- Close-fitting covers help to reduce evaporation of soap in the flow cell when it is not in use.
- Do not store flow cell for 1 week or longer without removing soap.
- Clean the flow cell and store dry.

Calibrator soap is a precisely concentrated and sterilized solution formulated to provide a clean, frictionless soap film bubble over the wide, dynamic range of the calibrator. The sterile feature of the soap is important to prevent residue buildup in the flow cell center tube, which could cause inaccurate readings. The use of any other soap is not recommended.

3.4 ELECTRONIC BUBBLE METER METHOD

Various electronic meters are available that mimic the performance of the bubble burette. Most use shortened bubble tubes and calculate this shortened tube length as though the bubble burette was the standard size. The protocol for using these meters is as follows:

- Connect the collection device, tubing, pump, and calibration apparatus.
- Visually inspect all Tygon® tubing connections.
- Wet the inside of the electronic flow cell with the soap solution by pushing on the button several times.
- Turn on the pump and adjust the pump rotameter, if available, to the appropriate flow rate.
- Press the button on the electronic bubble meter. Visually capture a single bubble and electronically time the bubble. The accompanying printer will automatically record the calibration reading in liters per minute.
- Repeat these steps until two readings are within 5%. If necessary, adjust the pump while it is still running.
- Repeat the procedures described above for all sampling pumps. The same cassette and filter may be used for calibrations involving the same sampling method. For sorbent tube sampling, however, the sorbent tube to be used must be used during the calibration.

Note: When calibrating with a bubble meter, cassette adapters can cause moderate to severe pressure drop at high-flow rates in the sampling train and affect the calibration result.

- If adapters are used for sampling, they should also be used for calibrating.
- **Caution:** Nylon adapters can restrict airflow due to plugging.
- Stainless-steel adapters are preferred.

3.5 DRY-FLOW CALIBRATION

With the advent of computer chip and microcircuitry technology, dry-flow calibration of instruments is now possible. Dry-flow calibrators measure the flow across near frictionless composite pistons (Figure 3.7). For some pumps, an adapter must be used between the pump and the dry-flow calibration instrument (Figure 3.8).

In addition to measuring flow rates, the calibration devices can also record time, date, employee names, pump ID, sample ID number, and other programmed information (Figure 3.9).

3.6 PRECISION ROTAMETER METHOD (SECONDARY)

The precision rotameter is a secondary calibration device. If used in place of a primary device such as a bubble meter, take care that any error introduced is minimal and noted (Figure 3.10).

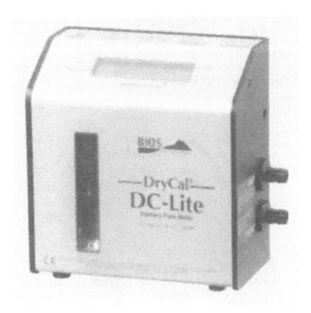

Figure 3.7 An electronic flow calibrator is available that does not use soap; instead it uses a graphite/carbon-composite piston that rises in the chamber like the bubble. (Bios International Corp.)

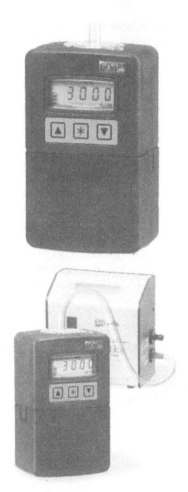

Figure 3.8 An electronic adapter is available to automatically calibrate an air sampling pump. (SKC—AirCheck® 2000 air sampling pump, CalCheck® Communicator with DC-Lite calibrator). (Bios International Corp.)

Figure 3.9 The electronic standardization and communication module provides time/date stamping; employee, pump, and sample ID numbers; volumetric flow rate readings; and readings at user-defined time intervals for flow stability testing. (Bios International Corp.)

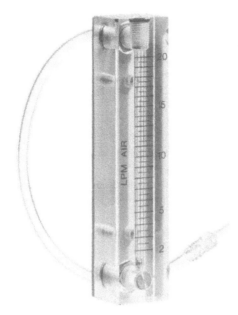

Figure 3.10 Secondary standard rotameters are used in the field to calibrate air-sampling pumps. Rotameters are calibrated to a primary standard. Flow rates must be corrected for standard temperature and pressure. (SKC)

3.6.1 Replacing the Bubble Meter with a Precision Rotameter

The precision rotameter may be used for calibrating the personal sampling pump in lieu of a bubble meter, provided it is

- Calibrated regularly, at least monthly, with an electronic bubble meter or a bubble meter.
- Disassembled, cleaned as necessary, and recalibrated. (It should be used with care to avoid dirt and dust contamination, which may affect the flow.)
- Not used at substantially different temperature and/or pressure levels than when the rotameter was calibrated against the primary source.
- Used in such a way that the pressure drop across the rotameters is minimized.

If altitude or temperature at the sampling site is substantially different from that at the calibration site, it is necessary to calibrate the precision rotameter at the sampling site.

3.7 SPAN GAS

Span gas of known concentration can be used to calibrate detectors. An example is the use of span gas in a PID.

A known concentration of a gas that readily ionizes in the PID energy lamp's electron voltage is drawn into the ionization chamber. Essentially this gas spans the distance between the anode and cathode sides of the ionization chamber. The detector must therefore detect this span and send an appropriate electrical signal to the readout device.

The electrical output to the detector readout for the PID is then adjusted. The reason for this adjustment or calibration is that with a known concentration of a gas that will ionize at a known voltage, the PID should detect and read out the same as the concentration listed on the compressed gas bottle.

Isobutylene has an IP of 9.8 eV and is the usual calibration gas of choice for PIDs. Historically, benzene was used (same IP of 9.8 eV); however, that use was discontinued due to benzene's carcinogenic and toxic effect potential. Remember that after the ionization cycle, the gases reform and are exhausted from the PID reaction chamber. The other exposure route occurred when the calibration gas was attached to the sampling train.

3.8 BUMP TESTING

Bump testing is used to check sensor operation. This sensor check is not a substitute for the calibration of sensors. However, bump testing can

- Provide an indication of sensor reliability under field conditions.
- Indicate when calibration is required.
- Test all sensors simultaneously.

Bump testing involves the use of a bump test gas cylinder filled with known concentrations of various gaseous challenge agents. These gases each provide a known percentage or parts per million component. The gas cylinder is attached to an instrument's inlet portal or diffusion grid, and the instrument's detector readouts are compared to the bump gas known concentration (Figure 3.11).

Figure 3.11   Bump testing needs to be performed regularly to ensure that real-time air monitoring instrumentation is functioning properly. (MSA)

CHAPTER 4

Statistical Analysis and Relevance

This chapter contains theoretical and real-world discussions about statistical analysis. It gives special emphasis to situations where knowledge of statistical relevance is a prerequisite for sampling adequacy. It also illustrates the difference between log-normal and normal distribution and parametric monitoring data.

4.1 DEFINITIONS

In statistical analysis and relevance, certain standard definitions are used. The following standard definitions and examples illustrate the basic concepts.

- **Accuracy:** Agreement of the measured value (i.e., empirical value) and the "true value" (i.e., accepted reference value) of the sample given valid sampling techniques, proper sample preparation, and reliable and accurate instrumentation and/or other procedures.

    Accuracy is often estimated by adding (or "spiking") known amounts of the target parameters. For asbestos quality assurance (QA) sampling, accuracy is evaluated by comparing analyses of duplicate samples that have been evaluated in proficiency in analytical testing (PAT) round robins (for air samples) or NIST National Voluntary Laboratory Accreditation Program (NVLAP) proficiency testing (for bulk and transmission electron microscopy [TEM]).

    Accuracy is a measure of the bias of the method and may be expressed as the difference between two values, a ratio, or the percentage difference.
- **Analysis:** Combination of sample preparation and evaluation.
- **Audit:** Systematic determination of the function or activity quality.
- **Bias:** Systemic error either inherent in the method or caused by measurement system artifacts or idiosyncrasies.
- **Blind sample:** Presented to the laboratory as indistinguishable from field samples (syn: performance audit samples). All field blanks are to be presented to the laboratory as blind samples for asbestos air samples.
- **Calibration:** Comparative procedure in which singular measurements are evaluated against an accepted group of measurements. The evaluation may be against a primary, intermediate, or secondary standard.

- **Calibration curve:** Range over which measurement can take place (syn: standard curve, multipoint calibration).
- **Calibration standard:** Instruments or other measurement techniques used to evaluate the measurement system. Ideally, these standards do not directly incorporate or use the target parameters to be measured.
- **Chain-of-custody:** Defined sample custody procedures that must be followed to document the transition from field collection to subsequent transfer sites (common carriers, laboratories, storage facilities, etc.).
- **Check standard:** Used to verify that the initial standard or calibration curve remains in effect. It ideally incorporates standard materials (syn: daily standard, calibration check or standard, reference standard, control standard, single point response factor, single point drift check).
- **Comparability:** Confidence with which one set of empirical data can be compared to another.
- **Completeness:** Amount of valid data obtained from a measurement system compared to the amount expected.
- **Data quality:** Totality of data parameters that identify ability to satisfy or represent a given condition; includes accuracy, precision, representativeness, and comparability.
- **Data reduction:** Using standard curves to interpret raw data.
- **Data validation:** Review process that compares a body of data against a set of criteria to provide data adequacy assurance given the data's intended use; includes data editing, screening, checking, auditing, verification, certification, and review.
- **Detection limit:** Minimum target parameter quantity that can be identified, i.e., distinguished from background or "zero" signal.
- **Double-blind sample:** When neither the composition nor identification of the sample is known to the analyst.
- **Duplicate sample (field or laboratory):** Sample divided into two portions, with both portions carried through the sample preparation process at the same time. For asbestos air samples, field duplicates are air samples collected at the same time as the compliance air samples, and lab duplicates are portions of one filter that are fixed and analyzed separately.
- **Environmentally related measurements:** Field or laboratory investigations that generate data involving chemical, physical, or biological parameters characteristic of the environment.
- **Field blanks:** Generated at the time of sampling, field blanks provide a check on contamination, starting with the sampling process and proceeding through the full analysis scheme. For asbestos fiber concentration sampling, field blanks are filter cassettes transported to the site and exposed to ambient conditions. The filter caps are removed from the filter cassettes; however, a vacuum air pump is not used to pull air across the filter cassettes. Thus, the cassettes are exposed to the environmental airstream of the surrounding environment outside the asbestos control area.
- **Good laboratory practices:** Performing a basic laboratory operation or activity so as not to influence data generation quality.
- **Instrument blank:** Used to obtain information on instrument aberration absence/presence. The measurement instrument is presented with materials normally within the instrument and cycled through the measurement sequence. The resulting signal is then defined as the baseline instrument signal level.
- **Internal standard:** A nontarget parameter added to samples just prior to measurement to monitor variation in sample introduction and stability and to normalize

data for quantitation purposes. Internal standards are not usually used in bulk (phase light microscopy [PLM]) or phase contrast microscopy (PCM); however, these standards may be applicable to TEM protocol.

- **Laboratory blank:** Prepared in the laboratory after receipt of samples from the field. These blanks are prepared using a material assumed not to contain the target parameter. Lab blanks for asbestos sampling are filter membranes obtained from filter cassettes that have been retained in the laboratory without removal of the filter cassette caps. The lab blank is a check on all the chemicals and reagents used in the method as well as the influence of the general laboratory environment (syn: analytical blank, system blank, method blank, process blank).

- **Measurement:** Creating quantitative data from a prepared sample.

- **Method check sample:** Prepared in the laboratory by spiking a clean reference matrix with known quantities of the target parameters. For asbestos air-sampling analysis, method check samples are previously prepared filters evaluated by separate analysts within the same laboratory. These duplicate analyses are defined in the National Institute of Occupational Safety & Health (NIOSH) 7400 method as quality assurance, and the acceptable statistical parameters are outlined therein.

- **Method detection limit:** Minimum quantity that a method (i.e., both sample preparation and target parameter measurement steps) can be expected to distinguish from background or "zero" signal. This limit takes into account losses during preparation and measurement and instrument sensitivity that may contribute to qualification or quantification of results. This limit does not apply to physical parameters (i.e., density, temperature).

- **Performance evaluation (PE) sample:** Sample with known "true" values that is presented to the laboratory as a "performance evaluation sample." These samples are biased by the analyst's knowledge of the intent of the sample. For asbestos air-sampling analysis, "true" value samples may be defined as the PAT samples with their inclusive statistical ranges.

- **Precision:** Measure of the reproducibility of a set of results obtained under similar conditions. Precision is determined by multiplicate analysis of samples, duplicates, replicates, or splits. Standard deviation is used as a measure of precision.

- **Procedure:** Systematic instructions and operations for using a method of sampling or measurement.

- **Proficiency sample:** Samples for which known composition values are available for accuracy comparisons. The composition values may be qualitative, quantitative, or statistical ranges of acceptable qualitative/quantitative results.

- **Quality assurance:** An orderly assemblage of management policies, objectives, principles, and general procedures by which a laboratory outlines the methods used to produce quality data. QA is an intralaboratory function. **Note:** The NIOSH 7400 method defines QA in terms of both intralaboratory and interlaboratory methods and/or sequencing. However, for the purposes of specified QA/quality control (QC) documents, interlaboratory methods are defined as QC.

- **Quality control:** Routine application of procedures used to develop prescribed performance standards in the monitoring and measurement of standards. QC is an interlaboratory function.

- **QC samples:** Analyzed concurrently with field samples to insure that analytical systems are operating properly, i.e., in control. These samples provide an estimate of the precision and accuracy of the sampling and analysis system. QC samples for asbestos sampling are sent between laboratories for interlaboratory comparisons of methodology and analytical proficiency.

- **Quality of method:** Degree to which the method functions free of systemic error, bias, and random error.
- **Quantitation limits:** Maximum and minimum levels or quantities of a reliably quantified target parameter. These limits are bounded by the standard curve limits and are generally related to standard curve data.
- **Reagent blank:** Used to identify contamination sources. These blanks incorporate specific reagents during sample preparation to identify lab blank contaminate sources (syn: dilution blank).
- **Recovery/percent recovery:** Generally used to report accuracy based on the measurement of target parameters, comparison of these concentrations, and correlating these measurements to the predicted amounts. Recovery in asbestos air sampling is the statistical percentage differential observed during accuracy evaluation (i.e., percentage difference in fiber concentrations).
- **Replicate samples (field or laboratory):** A sample is divided into two portions and is processed as two completely separate and nonparallel samples, i.e., prepared and analyzed at different times or by different people. Field replicates in asbestos air sampling are defined as filters that are obtained from two separate filter cassettes drawn separately or from a y-juncture. These are then transported, fixed, and analyzed separately. Lab replicates are taken from a singular filter, which is sectioned, fixed, and analyzed separately.
- **Representativeness:** Degree to which data accurately and precisely represent a parameter variation characteristic at a sampling point and portray an environmental condition.
- **Sample custody:** Verification and documentation procedure for the transfer of samples from the field to the laboratory, within the laboratory, and to the final storage or disposal destination.
- **Sample Operation Procedure (SOP):** Procedure adopted for repetitive use when performing a specific measurement or sampling operation.
- **Sample preparation:** Transformation of the sample into appropriate forms for transfer and/or measurement.
- **Sampling:** Removal of a process stream representative portion or a portion of a larger quantity of material for subsequent evaluation.
- **Sensitivity:** An instrument's detection limit given the minimum quantity of a target parameter that can be consistently identified, i.e., distinguished from background or "zero" signal by the instrument; ideally established using the materials that are used for standardization.
- **Split samples (field or laboratory):** A sample, divided into aliquots, that is sent to a different laboratory for preparation and measurement. These split samples may be replicates or duplicates that are then defined as splits when sent to another laboratory for QC analysis. For asbestos air samples this shipment involves either the shipping of capped filter cassettes (field split) or the shipping of fixed slides (lab split).
- **Standard materials:** Materials, such as mixtures of the target parameters at known concentration and purity, used to carry out standardizations. For asbestos air sampling these materials may be PAT samples.
- **Standardization:** Establishing a quantitative relationship between known target parameters input and instrument readout.
- **Target parameters:** Entity for which qualitative or quantitative information is desired.

- **Trip blanks:** Essentially field blanks that do not have the caps removed. These blanks provide insight into the contamination generated as a result of the shipping process. These blanks are generally not required for asbestos sample shipment.

4.2 EXAMPLE—OUTLINE OF BULK SAMPLING QA/QC PROCEDURE

Bulk sample analysis procedures are defined in the NIOSH method for PLM and the NIOSH 7402 method for TEM. Because bulk sample analysis is done by an independent laboratory off-site, this QA/QC document will not address bulk sampling as a field-verified procedure. The NVLAP is currently used to ascertain laboratory effectiveness. The contractor must provide proof that NVLAP certification is current for the laboratory designated to receive both the initial bulk samples and the 10% bulk sample duplicates.

The optical properties and ID of fibers are as follows:

- Determine 13 specific items for asbestos.
- ID other fibers with some optical data.
- ID matrix components.

In addition record

- Special procedures or solvents
- Sampling of layers
- Deviations from EPA test procedure

The 10% bulk sample duplicates are provided by physically dividing 10% of the samples collected. This division should occur concurrently with the collection of field samples. The rationale for splitting the samples at a time and location removed from the field collection site must be provided.

All field collection procedures, sample labeling, and transport and disposal procedures must be addressed in the QA/QC document. Provide the rationale for the classification and remediation of errors. The following is a sample of error classification:

Major Error

- False positive
- False negative (asbestos actually >1%)
- Incorrect asbestos type classification
- Analysis quantification in error by 25%

Minor Error

- Incorrect ID of tremolite or actinolite as another type of asbestos
- False negative (asbestos actually >1% or trace)
- Analysis quantification in error by 15%
- Incomplete lab data sheets (repeated omissions may equal major error)

Corrective Steps—Major Errors

- Take immediate action.

- Review documents for transcription errors.
- Review sample, especially matrix description.
- Reanalyze sample; submit to other labs for analysis.
- Check environment for contamination and out of calibration equipment.
- For misidentification of asbestos species review literature and reference samples.

Corrective Steps—Minor Errors

- One to two weeks or after monthly summaries
- Steps 1 through 4 above.
- Review to determine if error is systematic estimation bias, then retrain on estimation techniques and/or sample preparation techniques.
- Completed data sheets are required; frequent omissions should be considered a major error.
- Review sampling and stereomicroscope sample preparation procedures.
- To recalibrate estimation techniques, reanalyze known samples; review literature on estimation training.
- Review; retrain on reference samples, especially in problem matrix mixtures. Review specific problems like tremolite or crocidolite samples.
- Review lab's special sample preparations for Vinyl Asbestos Tile (VAT) or tar matrix materials. Practice on known materials including blanks.

4.3 EXAMPLE—OUTLINE OF THE NIOSH 7400 QA PROCEDURE

4.3.1 Precision: Laboratory Uses a Precision of 0.45

Current guide specifications give a precision of 0.45 as acceptable in calculation of the 95% upper confidence level (UCL). Using a precision of 0.45 implies that the standard deviation divided by the arithmetic mean gives a value of 0.45. This ratio is variously called the coefficient of variation (CV) or the SR in the NIOSH 7400 method. With the 90% confidence interval of mean count, which includes a subjective component of 0.45 plus the Poisson component, 0.45 precision implies that reproducibility of results is questionable. Thus, the use of the 0.45 value in the calculation of the UCL must be clearly identified. For compliance purposes the following equation is acceptable:

$$\text{Measured} + \text{(air quality) (0.45) (1.645) concentration (standard)}$$

The QA procedures given in the NIOSH 7400 method must be referenced in a discussion of the 0.45 precision value used. Detailed outlines of the intralaboratory procedures are not necessary. Proof of acceptable PAT participation (i.e., judged proficient for the target parameter in four successive round robins) as administered by the American Board of Industrial Hygiene (ABIH) must be provided.

4.3.2 Precision: Laboratory Uses a Precision SR that is Better Than 0.45

When a precision better than 0.45 is suggested, an outline of the NIOSH 7400 QA procedures used and a current CV curve must be provided in addition to the proof of acceptable PAT participation. The outline must include the following:

- Document the laboratory's precision for each counter for replicate fiber counts by using this procedure.
- Maintain as part of the QA program a set of reference slides to be used daily. These slides should consist of filter preparations including a range of loading and background dust levels from a variety of sources including both field and PAT samples.
  —Have the QA officer maintain custody of the reference slides and supply each counter with a minimum of one reference slide per workday. Change the labels on the reference slides periodically so that the counter does not become familiar with the samples.
  —From blind repeat counts on the reference slides, estimate the laboratory intra- and intercounter SR.
  —Determine separate values of relative standard deviation for each sample matrix analyzed in each of the following ranges. Maintain control charts for each of these data files.
    - 15 to 20 fibers in 100 graticule fields
    - >20 to 50 fibers in 100 graticule fields
    - >50 to 100 fibers in 100 graticule fields
    - 100 fibers in less than 100 graticule fields

Note: Certain sample matrices (e.g., asbestos cement) have been shown to give poor precision. Prepare and count field blanks along with the field samples. Report counts on each field blank.

Note 1: The identity of blank filters should be unknown to the counter until all counts have been completed.

Note 2: If a field blank yields greater than 7 fibers per graticule fields, report possible contamination of the samples.

- Perform blind recounts by the same counter on 10% of filters counted (slides relabeled by a person other than the counter). Use the following test to determine whether a pair of counts by the same counter on the same filter should be rejected because of possible bias:
- Discard the sample if the absolute value of the difference between the square roots of the two counts (in fiber/mm^2) exceeds 2.8 (x) 2 − 8 × SR, where x = the average of the square roots of the two fiber counts (in fiber/mm^2) and SR = one half the intracounter relative standard deviation for the appropriate count range (in fibers).

Note 1: Fiber counting is the measurement of randomly placed fibers that may be described by a Poisson distribution; therefore, a square root transformation of the fiber count data will result in approximately normally distributed data.

Note 2: If a pair of counts is rejected by this test, recount the remaining samples in the set and test the new counts against the first counts. Discard all rejected paired counts. It is not necessary to use this statistic on blank counts.

The analyst is a critical part of this analytical procedure. Care must be taken to provide a nonstressful and comfortable environment for fiber counting. An ergonomically designed chair should be used. With the microscope eyepiece situated at a comfortable height for viewing. External lighting should be set at an intensity level similar to the illumination level in the microscope to reduce eye fatigue. Counters should take 10–20 min breaks from the microscope every 1 to 2 hours to limit fatigue. During these breaks both eye and upper back/neck exercises should be performed to relieve strain.

Calculation of compliance uses the same equation with the substitution of a different value for 0.45. Use of this different value requires prior approval. All filters selected for the 10% recount should be randomly selected, rather than selecting every tenth filter.

4.3.3 Records to Be Kept in a QA/QC System

Records include:

- Sample logbook
- Chain-of-custody record

4.3.4 Field Monitoring Procedures—Air Sample

Collection of 10% duplicates as provided by specifications are not to be confused with the 10% duplication used for QA in the NIOSH 7400 method. The 10% duplicates provided by specifications are for QC interlaboratory determination of reliability. These duplicates are collected in the field using simultaneous collection devices or analysis preparatory techniques. Running two pumps on one individual is feasible for personal air sampling.

Running a Y shunt to two separate collection cassettes may be the technique used for area monitoring simultaneous collection. This method may also be feasible for personnel monitoring.

When neither type of collection alternative is feasible, duplication of prepared slides through duplicate filter media mounting is acceptable. The rationale and techniques for all prospective alternatives must be provided in the QA/QC document.

Caution: Field blanks must indeed be field blanks, not randomly prepared laboratory blanks. The purpose of field blanks is to access ambient air particulates hypothetically unassociated with abatement activities, but existing in the same general geographic location. Thus, field blanks must be filter-loaded cassettes that are open faced and stored outside the asbestos control area in an associated support zone. Occasionally, hygienists carry the field blanks in their pockets. This practice, if employed, must be documented, with associated persuasive rationale.

4.3.5 Calibration

All calibration techniques and schedules must be provided. Calibration of air-sampling devices may be accomplished using a primary standard (such as a bubble burette) or using a precision rotameter calibrated against a primary standard. The bubble gauge installed on the front vertical face of personal air-sampling pumps is not a reliable or precise calibration tool. Calibration must be done using dial settings versus primary standard timings. Rotameters offer an advantage in that their portability makes it possible to calibrate cassettes with the associated vacuum pump-generated airstream (or train) at the worksite.

Precision rotameters, while preferred, can be supplemented with standard rotameters. Techniques for rotameter use and calibration must be sequentially and clearly defined. The use of long-gauge range, 0–20 l/min, rotameters in the measure of personal air-sampling pumps with expected 2.5 l/min is prohibited due to lack of precision. Short-gauge range, 0–5 l/min rotameters, are allowed for use in the calibration of personal air sampling pumps. Example calibration curves and associated calculations must be provided.

Calibration of Phase Contrast Microscopy (PCM) microscopes used on-site must be completely defined. All NVLAP and American Conference Governmental Industry Hygienists (ACGIH) calibration procedures are enforceable for on-site activities. While not generally defined as a calibration technique, cleanliness and relative stability of the PCM microscopic location must be addressed. Sample calibration checklists used daily and weekly must be provided.

4.3.6 Negative Air Pressure

When negative air pressure is used in gross containment areas, monitoring criteria must be provided. Specifications should require continued real-time instrument monitoring independent of the HEPA vacuum system monitors. Monitors must be located outside the containment area and removed from the effluent HEPA vacuum airstream. Appropriate monitoring checklists and sample direct readout tapes must be provided. The readout tapes and associated calculations, if needed, can either be generically presented or be samples from previous monitoring efforts.

4.3.7 Compressor

In the event that compressed air is used on-site, certification of Grade D breathing air must be provided. If a filter bank is used in conjunction with an oil-lubricated compressor, monitoring of the filter bank is required. This monitoring includes carbon monoxide, temperature, oil breakthrough, and air pressure criteria. In addition to audible alarms and escape air, visual monitoring of the compressor filter bank status is necessary. Provide sample monitoring check sheets. Certain specifications call for the use of colorimetric tubes to certify continued Grade D breathing air supply. Provide sample monitoring check sheets that clearly indicate sampling intervals.

4.3.8 Recordkeeping and Sample Storage

Recordkeeping priorities and samples of format used must be provided. Records include documentation of air sampling and air sample analysis, bulk sampling and bulk sample analysis, and negative air pressure maintenance. Personnel records include résumés, certifications, medical surveillance, and training. Environmental records detail work sequencing and ambient conditions. All specified documents must be accessible and maintained as per specifications. Sample storage and accessibility must be discussed.

The following equipment and documentation lists are examples of those that should be appraised for asbestos bulk sampling. Other laboratory protocols will require similar lists.

Laboratory Procedures

- Logbook
- Calibration of refractive index oils
- Daily microscope alignment
- Daily microscope calibration check
- Daily microscope contamination check
- Equipment maintenance
- Equipment calibration
- Personnel records, including hierarchy, training, certification, and job descriptions

- Monthly records of each analyst
- QA and QC results for their work
- Proficiency results (PAT)
- Precision and accuracy ratings, including explanation of rating protocols

QA Logbook

- Samples
- Results
- Discrepancies
- Analysis repeats (minimum 10%)
- Intralaboratory analysis of proficiency samples
- Intralaboratory analysis of duplicates and replicates
- Blank analysis
- Summary of results from each analyst
- Summary of results for the laboratory

QC Logbook

- Deficiency corrections
- Samples
- Results
- Discrepancies
- Frequency of duplicate/replicate analysis per total samples
- Interlaboratory analysis of proficiency samples
- Timing of QC analysis
- Same day, next shift, next day
- Monthly proficiency samples, WULAP samples in-house or past EPA asbestos bulk sample analysis, QA program samples, blanks, and contamination samples

Interlaboratory Analyses (summary of results for the laboratory)

- Outliers
- Interlaboratory analysis schedules
- Time, including expected turnaround time
- Labs participating
- Contamination testing and control logbook
- Lab data sheet/notebook
- Analysis report sheet
- QA manual revision documentation
- Training procedures for staff
- Analysis error correction correspondence

4.4 SAMPLING AND ANALYTICAL ERRORS

When an employee's personal exposure or the area exposure is sampled and the results analyzed, the measured exposure will rarely be the same as the true exposure. This variation is due to sampling and analytical errors or SAEs. The total error depends on the combined effects of the contributing errors inherent in sampling, analysis, and pump flow.

Error factors determined by statistical methods shall be incorporated into the sample results to obtain:

- The lowest value that the true exposure could be (with a given degree of confidence)
- The highest value the true exposure could be (also with some degree of confidence)

The lower value is called the lower confidence limit (LCL), and the upper value is the UCL. These confidence limits are one sided, since the only concern is with being confident that the true exposure is on one side of the Permissible Exposure Limit (PEL).

4.4.1 Determining SAEs

SAEs that provide a 95% confidence limit have been developed and are listed on the OSHA-91B report form (most current SAEs). If there is no SAE listed in the OSHA-91B for a specific substance, call the laboratory. If using detector tubes or direct-reading instruments, use the SAEs provided by the manufacturer.

4.4.2 Environmental Variables

Environmental variables generally far exceed SAE. Samples taken on a given day are used by OSHA to determine compliance with PELs. However, where samples are taken over a period of time (as is the practice of some employers), the industrial hygienist should review the long-term pattern and compare it with the results. When OSHA's samples fit the long-term pattern, it helps to support the compliance determination. When OSHA's results differ substantially from the historical pattern, the industrial hygienist should investigate the cause of this difference and perhaps conduct additional sampling.

4.4.3 Confidence Limits

One-sided confidence limits can be used by OSHA to classify the measured exposure into the following categories.

95% Confident That the Employer Is in Compliance

- Measured exposure results do not exceed the PEL.
- UCL of that exposure does not exceed the PEL.

95% Confident That the Employer Is NOT in Compliance

- Measured exposure results do exceed the PEL.
- LCL of that exposure does exceed the PEL.

95% Confident That the Employer Is in Compliance = Possible Overexposure

- *Measured exposure results do not exceed the PEL.*
- UCL of that exposure does exceed the PEL.

**NOT 95% Confident That the Employer Is NOT in Compliance =
Possible Overexposure**

- Measured exposure results do exceed the PEL.
- LCL of that exposure does not exceed the PEL.

A violation is not established if the measured exposure is in the "possible over-exposure" region. It should be noted that the closer the LCL comes to exceeding the PEL, the more probable it becomes that the employer is in noncompliance.

If measured results are in this region, the industrial hygienist should consider further sampling, taking into consideration the seriousness of the hazard, pending citations, and how close the LCL is to exceeding the PEL.

If further sampling is not conducted, or if additional measured exposures still fall into the "possible overexposure" region, the industrial hygienist should carefully explain to the employer and employee that the exposed employee(s) may be overexposed, but that there were insufficient data to document noncompliance. The employer should be encouraged to voluntarily reduce the exposure and/or to conduct further sampling to assure that exposures are not in excess of the standard.

4.5 SAMPLING METHODS

The LCL and UCL are calculated differently depending on the type of sampling method used. Sampling methods can be classified into one of three categories:

4.5.1 Full-Period, Continuous Single Sampling

Full-period, continuous single sampling is defined as sampling over the entire sample period with only one sample. The sampling may be for a full-shift sample or for a short period ceiling determination.

4.5.2 Full-Period, Consecutive Sampling

Full-period, consecutive sampling is defined as sampling using multiple consecutive samples of equal or unequal time duration that, if combined, equal the total duration of the sample period. An example would be taking four 2-hour charcoal tube samples.

There are several advantages to this type of sampling. If a single sample is lost during the sampling period due to pump failure, gross contamination, etc., at least some data will have been collected to evaluate the exposure. The use of multiple samples will result in slightly lower SAE. The collection of several samples leads to conclusions concerning the manner in which differing segments of the workday affect overall exposure.

4.5.3 Grab Sampling

Grab sampling is defined as collecting a number of short-term samples at various times during the sample period that, when combined, provide an estimate of exposure over the total period. Common examples include the use of detector tubes or direct-reading instrumentation (with intermittent readings).

4.6 CALCULATIONS

If the initial and final calibration flow rates are different, a volume calculated using the highest flow rate should be reported to the laboratory. If compliance is not established using the lowest flow rate, further sampling should be considered.

Sampling is generally conducted at approximately the same temperature and barometric pressure as calibration, in which case no correction for temperature and pressure is required, and the sample volume reported to the laboratory is the volume actually measured. Where sampling is conducted at a substantially different temperature or pressure than calibration, an adjustment to the measured air volume may be required depending on sampling pump used, in order to obtain the actual air volume sampled. The actual volume of air sampled at the sampling site is reported and used in all calculations.

For particulates the laboratory reports milligrams per cubic meter of contaminant using the actual volume of air collected at the sampling site. This value can be compared directly to OSHA Toxic and Hazardous Substances Standards (e.g., 29 CFR 1910.1000).

The laboratory normally does not measure concentrations of gases and vapors directly in parts per million. Rather, most analytical techniques determine the total weight of contaminant in a collection medium. Using the air volume provided by the industrial hygienist, the lab calculates the concentration in milligrams per cubic meter and converts this to parts per million at 25°C and 760 mmHg. This result is to be compared with the PEL without adjustment for temperature and pressure at the sampling site.

$$ppm(NTP) = mg/m^3(24.45)/(Mwt)$$

where

- 24.45 = molar volume at 25°C (298 K) and 760 mmHg
- Mwt = molecular weight
- NTP = normal temperature and pressure at 25°C and 760 mmHg

If it is necessary to know the actual concentration in parts per million at the sampling site, it can be derived from the laboratory results reported by using the following equation:

$$ppm(PT) = ppm(NTP) (760)/(P) (T)/(298)$$

where

- P = sampling site pressure (mmHg)
- T = sampling site temperature (K)
- 298 = temperature in K

Since $ppm(NTP) = mg/m^3 (24.45)/(Mwt)$
$ppm(PT) = mg/m^3 \times 24.45/Mwt \times 760/P \times T/298$

Note: When a laboratory result is reported as milligrams per cubic meter contaminant, concentrations expressed as parts per million (PT) cannot be compared directly to the standards table without converting to NTP.

Note: Barometric pressure can be obtained by calling the local weather station or airport and requesting the unadjusted barometric pressure. If these sources are not available, then a *rule of thumb* is for every 1000 ft increase in elevation, the barometric pressure decreases by 1 in.Hg.

4.6.1 Calculation Method for a Full-Period, Continuous Single Sample

Obtain the full-period sampling result (value X), the PEL, and the SAE. The SAE can be obtained from the OSHA Chemical Information Manual. Divide X by the PEL to determine Y, the standardized concentration, that is, $Y = X/PEL$.

Compute the UCL (95%) as follows: $UCL(95\%) = Y + SAE$

Compute the LCL (95%) as follows: $LCL(95\%) = Y - SAE$

Classify the exposure according to the following classification system:

- If the UCL ≤ 1.0, a violation does not exist.
- If the LCL ≤ 1.0 and the UCL > 1, classify as possible overexposure.
- If the LCL > 1.0, a violation exists.

4.6.2 Sample Calculation for a Full-Period, Continuous Single Sample

A single fiberglass filter and personal pump were used to sample for carbaryl for a 7-h period. The industrial hygienist was able to document that the exposure during the remaining unsampled 0.5 h of the 8-h shift would equal the exposure measured during the 7-h period. The laboratory reported 6.07 mg/m³. The SAE for this method is 0.23. The PEL is 5.0 mg/m³.

Step 1. Calculate the standardized concentration.
$Y = 6.07/5.0 = 1.21$

Step 2. Calculate confidence limits.
$LCL = 1.21 - 0.23 = 0.98$
Since the LCL does not exceed 1.0, noncompliance is not established. The UCL is then calculated: $UCL = 1.21 + 0.23 = 1.44$

Step 3. Classify the exposure.
Since the LCL ≤ 1.0 and the UCL > 1.0, classify as possible overexposure.

4.6.3 Calculation Method for a Full-Period Consecutive Sampling

The use of multiple consecutive samples will result in slightly lower SAEs than the use of one continuous sample because the inherent errors tend to partially cancel each other. The mathematical calculations, however, are somewhat more complicated. If preferred, the industrial hygienist may first determine if compliance or noncompliance can be established using the calculation method noted for a full-period, continuous, single-sample measurement. If results fall into the "possible overexposure" region using this method, a more exact calculation should be performed as follows.

Compile $X(1)$, $X(2)$ . . . , $X(n)$, and the n consecutive concentrations on one workshift.

Compile their time durations, $T(1)$, $T(2)$, . . . , $T(n)$.
Compile the SAE.
Compute the TWA exposure.
Divide the TWA exposure by the PEL to find Y, the standardized average (TWA/PEL).
Compute the UCL (95%) as follows: UCL (95%) = Y + SAE (Equation E).
Compute the LCL (95%) as follows: LCL (95%) = Y − SAE (Equation F).

Classify the exposure according to the following classification system:

- If the UCL ≤ 1.0, a violation does not exist.
- If the LCL ≤ 1.0, and the UCL > 1, classify as possible overexposure.
- If the LCL > 1.0, a violation exists.

When the LCL ≤ 1.0 and the UCL > 1.0, the results are in the "possible overexposure" region, and the industrial hygienist must analyze the data using the more exact calculation for full-period consecutive sampling, as follows:

$$LCL = \frac{Y - SAE(T_1^2 X_1^2 + T_2^2 X_2^2 \ldots + T_n^2 X_n^2)^{1/2}}{PEL(T_1 + T_2 + T_n)}$$

4.6.4 Sample Calculation for Full-Period Consecutive Sampling

Two consecutive samples were taken for carbaryl instead of one continuous sample, and the following results were obtained:

Sample	A	B
Sampling rate (L/min)	2.0	2.0
Time (min)	240.0	210.0
Volume (L)	480.0	420.0
Weight (mg)	3.005	2.457
Concentration (mg/m³)	6.26	5.85

The SAE for carbaryl is 0.23.

Step 1. Calculate the UCL and the LCL from the sampling and analytical results:
TWA = (6.26 mg/m³ × 240 min + (5.85 mg/m³) × 210 min
450 min = 6.07 mg/m³
Y = 6.07 mg/m³/PEL = 6.07/5.0 = 1.21
Assuming a continuous sample: LCL = 1.21 − 0.23 = 0.98
UCL = 1.21 + 0.23 = 1.44
Step 2. Since the LCL < 1.0 and the UCL > 1.0, the results are in the possible overexposure region, and the industrial hygienist must analyze the data using a more exact calculation for full-period consecutive sampling. If the LCL > 1.0, a violation is established.

4.7 GRAB SAMPLING

If a series of grab samples (e.g., detector tubes) is used to determine compliance with either an 8-h TWA limit or a ceiling limit, consult with an industrial hygienist (ARA) regarding sampling strategy and the necessary statistical treatment of the results obtained.

4.8 SAES—EXPOSURE TO CHEMICAL MIXTURES

Often an employee is simultaneously exposed to a variety of chemical substances in the workplace. Synergistic toxic effects on a target organ are common for such exposures in many construction and manufacturing processes. This type of exposure can also occur when impurities are present in single chemical operations. New PELs for mixtures, such as the recent welding fume standard ($5 \, mg/m^3$), addresses the complex problem of synergistic exposures and their health effects. In addition 29 CFR 1910.1000 contains a computational approach to assess exposure to a mixture. This calculation should be used when components in the mixture pose a synergistic threat to worker health.

Whether using a single standard or the mixture calculation, the SAE of the individual constituents must be considered before arriving at a final compliance decision. These SAEs can be pooled and weighted to give a control limit for the synergistic mixture. To illustrate this control limit, the following example using the mixture calculation is shown. The mixture calculation is expressed as:

$$E_m = (C_1/L_1 + C_2/L_2) + \ldots C_n/L_n)$$

where

- E_m = equivalent exposure for a mixture (E_m should be ≤ 1 for compliance)
- C = concentration of a particular substance
- L = PEL

For example, to calculate exposure to three different, but synergistic substances:

Material	8-h exposure	8-h TWA PEL (ppm)	SAE
Substance 1	500	1000	0.089
Substance 2	80	200	0.11
Substance 3	70	200	0.18

Using Equation I: $E_m = 500/1000 + 80/200 + 70/200 = 1.25$

Since $E_m > 1$, an overexposure appears to have occurred; however, the SAE for each substance also needs to be considered:

- Exposure ratio (for each substance): $Y_n = C_n/L_n$
- Ratio to total exposure: $R_1 = Y_1/E_{m1} \ldots R_n = Y_n/E_m$
- The SAEs (95% confidence) of the substance comprising the mixture can be pooled by:

$$(RS_t^2) = [(R_1^2)(SAE_1^2) + (R_2^2)(SAE_2^2) + \ldots (R_n^2)(SAE_n^2)]$$

- The mixture control limit (CL) is equivalent to $1 + RS_t$.

  —If $E_m \leq CL_1$, then an overexposure has not been established at the 95% confidence level; further sampling may be necessary.
  —If $E_m > 1$ and $E_m > CL_1$, then an overexposure has occurred (95% confidence).

- Using the mixture data above:

$$Y_1 = 500/1000 \qquad Y_2 = 80/200 \qquad Y_3 = 70/200$$
$$Y_1 = 0.5 \qquad Y_2 = 0.4 \qquad Y_3 = 0.35$$
$$R_1 = Y_1/E_m = 0.4 \qquad R_2 = 0.32 \qquad R_3 = 0.28$$

- $(RS_t)^2 = (0.4^2)(0.089^2) + (0.32^2)(0.11^2) + (0.28^2)(0.18^2)$
- $RS_t = [(RS_t)^2)](1/2) = 0.071$
- $CL = 1 + RS_t = 1.071$
- $E_m = 1.25$

Therefore $E_m > CL$ and an overexposure has occurred within 95% confidence limits. This calculation is also used when considering a standard such as the one for total welding fumes.

CHAPTER 5

Chemical Risk Assessment

Real-world examples portray the decision logic needed to conduct chemical sampling when assessing risk. This chapter includes a troubleshooting section/checklist to assist samplers in either choosing a consultant or appraising in-house sampling methodology.

Chemical risk assessment is a twofold process. One part occurs off-site as known chemical information is assessed and calculations based on accepted formulas are done. The EPA baseline risk assessments (BLRAs), screening assessments, and remedial investigation studies rely on a body of knowledge accumulated over the last 20 years. Decisions about supportive air monitoring and actual on-site monitoring required during sampling events should also be made at this time.

The second stage is the actual accumulation of data during which workers must be protected against airborne hazards, including those resulting from their sampling efforts, including disturbance of the on-site medium (soil, water). Decision-making concerning personal protection and engineering controls may require air monitoring of personnel, area of influence, and the site perimeter.

In order to understand the context under which air monitoring protocols are developed, an understanding of chemical risk assessment for these sites is necessary. Keep in mind that the term *site* is an all inclusive one for this section and may include active industrial and/or construction sites.

5.1 BASELINE RISK ASSESSMENT

Monitoring to determine chemical risk may lead to a BLRA consistent with U.S. EPA Comprehensive Environmental Resource Conservation Liability Act (CERCLA) guidance documents that address:

- Potentially contaminated groundwater
- Surface water runoff, sediment, and river area
- Soils

The results of the BLRA may be used to

- Prioritize the need for site remediation or abatement activities.
- *Provide the basis for quantification of remedial objectives.*
- Assist in planning objectives to minimize risk.

5.2 CONCEPTUAL SITE MODEL

The first step in developing a BLRA is to provide a conceptual site model that has been developed to evaluate source areas, migration pathways, and possible exposure points for receptors. Migration pathways are potential conduits for contaminants to reach on-site and off-site receptors. This model is then used to determine the medium that needs to be sampled.

5.2.1 Source Areas

The source areas are limited to the areas delineated by the model. Source areas such as soil areas, bodies of water, and air emissions are areas for concern. Soils in particular may be primary and secondary source areas—primary as particulates that may lead to ingestion or dermal hazard, and secondary as soils that may be dispersed in the airstream through on-site activities.

5.2.2 Possible Receptors

In accordance with the EPA standard default exposure factors (SDEFs) guidance, construction worker, commercial/industrial, and recreational populations are considered possible receptors in the human health evaluation. Sampling efforts to quantify and qualify the potential exposure pathways will focus on predictive sampling of the medium of concern, including surface soil, subsurface soil, groundwater, and air. The source areas are defined as the current industrial use soil contamination.

5.3 CHEMICALS OF POTENTIAL CONCERN

Groundwater and soil analytical data from samples collected from the site area are evaluated for preliminary determination of chemicals of potential concern (COPCs). Samples are from known hot spots as a worst-case scenario for **soils** and are evaluated using log-normal distribution, Kriging, or Monte Carlo analysis. Ground- and surface-water samples are collected (20 samples per location) and are evaluated using normal distribution assumptions.

The frequency of detection of chemicals in each medium is calculated as the number of total detections out of the total number of analytical samples for each medium. Duplicate sample data will not considered in this calculation.

The positively identified chemicals for each medium are reviewed to identify chemicals that could potentially result in adverse health effects in humans or in adverse environmental effects. Detected concentrations are compared to potential applicable or relevant and appropriate requirements (ARARs). ARARs include Safe Drinking Water Act (SDWA) maximum contaminant levels (MCLs), lifetime health advisory levels (HALs), and U.S. EPA Region VII preliminary remediation goals (PRGs). If detected concentrations are significantly below health-based ARARs or if compounds are known to be nontoxic, they are eliminated from further consideration.

5.4 HUMAN HEALTH BLRA CRITERIA

To understand the rationale for selecting sample sites, the sampler must have a basic understanding of risk assessment criteria. The following components of a risk assessment are normally part of all risk assessment discussions—even if those discussions are ultimately negative declarations—"we do not have to worry about that."

5.5 TOXICITY ASSESSMENT

The toxicity assessment weighs available evidence that COPCs may cause adverse health effects in exposed humans or other biota. The assessment estimates the extent of potential chemical exposure and the increased likelihood and/or severity of adverse effects. The toxicity assessment at the site location is accomplished in two steps:

1. Chemical hazard identification determines whether potential chemical exposure causes an increased incidence or severity of an adverse health or environmental effect. Toxicological data for COPCs are reviewed, and toxicological profiles are prepared for the COPCs.
2. The dose-response evaluation consists of quantitatively evaluating the toxicity information. Then the relationship between the chemical dose and the resultant incidence/severity of adverse health or environmental effects are reviewed.

The risk characterization portion of the BLRA estimates the likelihood of adverse effects occurring under the exposure scenarios. For the chemicals identified (except lead) the toxicity values have been derived by the U.S. EPA and are summarized in the Integrated Risk Information System (IRIS) database.

The EPA has not developed a reference dose (RfD) or carcinogenic potency slope factor (SF) for elemental lead. The EPA considers lead a special case because lead is ubiquitous in all media; therefore, human exposure comes from multiple sources. Thus most people would exceed an RfD level under "background" conditions. The EPA Office of Solid Waste and Emergency Response (OSWER) has released a directive on BLRA and cleanup of residential soil lead. This directive recommends that soil lead levels less than 400 ppm are considered safe for residential use. The EPA action level (SDWA) for lead in drinking water is 15 μg/l. An RfD, 1×10^{-7} mg/kg/day, has been provided for tetraethyl lead, formerly a common additive for gasolines in the U.S. The current accepted site usage levels for lead are evaluated in concert with the surrounding area soil levels.

For noncarcinogens, an RfD is established by the EPA as a result of hazard identification and dose-response evaluation. The RfD is an estimate of an exposure level judged likely to be without an appreciable risk of adverse health effects over a specified time of exposure. A critical study (or studies) in which a dose causing an adverse effect is identified at a lower level of exposure as either having no effect or minimal effect—yields the RfD. Chronic RfDs reflect a level of exposure that would not result in adverse effects when experienced for 7 years to a lifetime (Baseline Risk Assessment Guidance for Superfund [RAGS] Part A). If the RfD is expressed as an administered dose, dermal toxicity values are derived by adjusting the oral toxicity value (i.e., multiplying by a chemical absorption factor).

The noncarcinogenic effects potential is evaluated by comparing the chemical intake with an RfD. This ratio is referred to as a hazard quotient (HQ). The HQ assumes that below a certain level of exposure, even sensitive populations will not experience adverse effects. Based on this assumption, if exposure is equivalent to or less than the RfD and the HQ is 1.0 or less, a hazard is not likely to exist. If the HQ exceeds 1.0, a hazard may exist.

For carcinogens a cancer SF and a weight-of-evidence classification are the toxicity values used in the characterization of potential human carcinogenic risks. The relationship relating exposure level to the probability of developing cancer (i.e., the incidence) is expressed as a cancer SF.

The SF is a plausible upper-bound estimate of the probability of developing cancer per unit intake of a chemical over a lifetime. SFs are expressed as $(mg/kg\text{-}day)^{-1}$. If the SF is expressed as an administered dose, a dermal toxicity value is derived by adjusting the oral toxicity value (i.e., dividing by the chemical absorption factor).

The potential for carcinogenic effects is estimated by multiplying a chemical's SF by the lifetime average daily chemical intake. Exposure to carcinogens resulting in an increased carcinogenic risk of 1×10^{-6} or greater may be cause for potential concern and may indicate the need for remedial action.

The risk of developing cancer as a result of exposure to carcinogens can also be expressed as a unit risk. This toxicity value represents a risk per unit concentration in the particular medium contacted. Unit risks reflect risks resulting from continuous lifetime exposures. Unit risks are expressed as $(\mu g/m^3)^{-1}$ for inhalation exposures or $(\mu g/l)^{-1}$ for oral exposures. A unit risk in the range of 1×10^{-1} to 1×10^{-6} implies that an individual has between a 1 in 10,000 and a 1 in 1,000,000 chance of developing cancer in excess of a background incidence if exposed to $1\ \mu g/m^3$ air or $1\ \mu g/l$ water of a carcinogenic chemical for a lifetime.

The carcinogenic potential of a chemical is classified into one of the following groups according to the weight of evidence from epidemiological and animal studies:

- A—human carcinogen
- B1—probable human carcinogen (limited evidence of carcinogenicity in humans)
- B2—probable human carcinogen (sufficient evidence in animals with inadequate evidence in humans)
- C—possible human carcinogen (limited evidence of carcinogenicity in animals or lack of human data)
- D—not classifiable as a human carcinogen
- E—evidence of noncarcinogenicity in humans

5.6 TOXICOLOGICAL PROFILES

The Agency for Toxic Substances and Disease Registry (ATSDR) data and IRIS information are used to prepare toxicological profiles for COPCs. The toxicological profiles will include specific toxicological information (e.g., toxicological effects, target organs, critical effect).

5.7 UNCERTAINTIES RELATED TO TOXICITY INFORMATION

The RfDs established for COPCs are a major source of uncertainty in a BLRA. The RfD is the estimate of daily exposure likely to be without an appreciable risk of deleterious effects during a lifetime. The RfD is derived by the application of uncertainty factors to selected exposure levels identified in animal or human studies. Identified exposure levels are divided by these uncertainty factors to assure that the RfD will not be overestimated.

For example, an uncertainty factor of 10 is used to account for variations in human sensitivity when using data from valid human studies involving long-term exposure of average, healthy subjects. Additional uncertainty factors of 10 are applied to account for uncertainties in extrapolating from observation of toxicity in animals to predicted toxicity in humans, to account for uncertainties in identifying threshold dose from experimental data, and to account for uncertainties in extrapolating from subchronic to chronic studies. Any additional uncertainty factor or modifying factor ranging from >0 to ≤10 may be applied to reflect professional assessment of other uncertainties that may exist in the toxicity database for a specific compound.

Considerable uncertainties are involved in identifying whether or not a compound is a likely potential human carcinogen and at what level of exposure an increased risk of cancer may exist.

Uncertainties in quantifying the exposure level that may result in elevated carcinogenic risk for specific compounds are compensated for by using the 95% UCL of the estimated slope. This slope refers to the line that relates chemical exposure to the probability of developing cancer—thus the term *slope factor* for carcinogens. Using the 95% UCL is a statistical path to assure that the actual SF is highly unlikely to be greater than the SF listed for that chemical. These dose-response assumptions provide an upper, but plausible, estimate of the limit of risk when the SF is used to estimate risk associated with an estimated level of exposure.

5.8 POTENTIALLY EXPOSED POPULATIONS

For the future land-use scenario the assumption is that the site property will remain a site. The population to be considered includes on-site workers, site visitors, and possible future construction workers. The population may be exposed to surface soil, subsurface soil, and groundwater through inhalation, ingestion, and dermal contact.

The monitoring results of the well survey efforts are used to determine completed exposure pathways. The site does not have on-site residential or adjacent residential property; therefore, the soil exposure pathway is not applicable (i.e., complete) for on-site or adjacent sites.

Recreational receptors to be evaluated will include site trespassers and users of the site.

5.8.1 Exposure Pathways

The exposure pathway is the course a chemical takes from the source to the exposed individual. The exposure pathway is characterized by the source (the contaminated medium), the mechanism of release, a retention or transport medium, a point of potential human contact with a contaminated medium, and an exposure route at the point of contact (i.e., ingestion). An exposure pathway is complete only when each of these elements is present. Air monitoring may be used to supplement theoretical calculations of air dispersion pathways.

5.8.2 Sources

Surface soil and subsurface hydrogeological formations (i.e., soil/groundwater system) that have been affected by the site's former usage are the source areas that come under consideration.

5.9 ENVIRONMENTAL FATE AND TRANSPORT OF COPCs

The inorganic contaminants identified at the site may not be generally considered mobile in the soil/groundwater system. Mobility is affected by soil pH and groundwater for aqueous transport. Factors influencing dust generation and movement from the soil surfaces to the air migration pathway must also be considered. Inorganic contaminant mobility through storm water runoff and surface water transport are evaluated. Existing site conditions are evaluated to estimate the migration of contaminants to subsurface soil, groundwater, and surface water, and air.

5.10 EXPOSURE POINTS AND EXPOSURE ROUTES

Only complete exposure pathways involving current or future contact with contaminated media are cause for concern. Exposure pathways, in that the potentially exposed population is not likely to experience significant contaminated medium contact or the environmental medium contacted is not significantly contaminated, will **not** be considered. Therefore, before determining that the pathways are complete and that the possible receptors are likely to be exposed at significant levels, site-specific information is gathered. This information will also be used to develop site-specific exposure parameters. An interim BLRA deliverable is prepared to propose the exposure scenarios that are used in the final BLRA.

5.11 COMPLETE EXPOSURE PATHWAYS EVALUATED

A well survey is conducted to determine if any down-gradient domestic wells at the site may be affected. Based on the proximity of the wells, their depth, construction, and use, the potential for these wells to be affected by the site are determined. Irrigation wells, if within the site's vicinity, are a potential exposure pathway vector to the fields that are irrigated.

For the industrial exposure scenario, on-site workers and site visitors are considered for exposure to contamination through surface soil, subsurface soil, and air. The recreational exposure scenarios are considered for site trespassers and visitors to the site.

5.12 ECOLOGICAL RISK ASSESSMENT

Pathways for terrestrial and aquatic environmental receptors regarding exposure to subsurface soil, surface soil, sediment, surface water, and air are considered in the ecological component of the risk assessment. The ecological component of the BLRA characterization includes the following:

- Identification of habitats and predominant species occurring at the site including the river areas
- Selection of representative species that may have the greatest exposure based on feeding habits, habitat usage, and exposure duration (Receptors of concern will also be evaluated based on available published eco-toxicological data.)
- Verification that contaminants of potential ecological concern are limited to heavy metals

- Research into toxicological reference values in published material (Ambient water quality criteria [AWQC] are used for contaminants.)
- Evaluation of environmental data to adequately estimate existing and potential future ecological risks
- Development of a sampling and analysis plan (SAP) for data collection to evaluate ecological exposure
- Determination whether background data are required or if other information such as surface water hardness is necessary to evaluate toxicity
- Quantification of exposure by estimating the magnitude and rate of exposure for receptors of concern. (This quantification will include evaluation of contaminant concentrations, bioavailability, bioaccumulation, bioconcentration, and biomagnification potential; feeding rates; habitat usage; and food chain considerations.)
- Review of the results of the toxicity and exposure assessments
- Comparisons of exposure concentrations to appropriate toxicological reference values to complete a risk characterization (This characterization will include an evaluation of the spatial distribution of contamination with regard to ecological receptors.)

Decisions on whether an ecological impact exists are based on this risk characterization. Sampling to determine contaminant exposure potential then follows the same criteria as for the human health risk assessment, except that sampling routines may be specialized to deal with species-specific exposures. An example would be the exposure of burrowing animals to soils through ingestion, inhalation, and dermal contact.

5.13 DATA EVALUATION AND DATA GAPS

Existing data for concentrations of contaminants in the medium of concern—surface soil, subsurface soil, groundwater, and air—are evaluated. This evaluation will determine if data are adequate to estimate exposure point concentrations and to evaluate contaminant migration and toxicity. Existing data will also be evaluated to determine if data are of adequate quality for use in a risk assessment according to the methods specified in the U.S. EPA guidance, *Data Usability in Risk Assessments.*

If additional environmental data are necessary to complete the BLRA, an SAP is developed. In addition to chemical data, collection of water quality data (i.e., hardness) may be specified in the SAP because the toxicity and mobility of metals in surface water is hardness dependent.

At this time an evaluation of existing data indicates that on-site groundwater data are available. Current groundwater data and new empirical data are evaluated for background and down-gradient information. The arithmetic means and 95% UCLs of the mean are calculated for COPCs. For the groundwater pathways a well survey is conducted within 1 mile of the site to verify the locations and uses of all wells.

If recontouring of the soil surface has occurred to direct storm water runoff toward collection points, additional surface soil samples (0–2 in.) may be necessary for the analysis of COPCs in the soil ingestion and air pathways.

Data collection points are designed to identify hot spots and to calculate average concentrations over the entire site and in the areas of concern. Soil pH must also be measured because soil pH influences metals transport.

Surface water data for up-gradient and down-gradient points and sample collection points for this data are identified. No surface water or sediment data from on-site drainages, or drainage pathways from the site, may be available to provide information on the extent of contaminant migration. Sediment data from the surface water pathway provide information on whether the rivers or standing bodies of water subject to runoff are another source of contamination to other surface waters. Collect surface water and sediment samples from these areas.

The level of generation of dust at the site is required information. This information may be collected via real-time particulate measurements and/or by collecting samples for laboratory analysis. The level of dust generation is assumed to vary significantly with time and climatic data. The collection of dust generation data is planned carefully so that dust generation is neither underestimated or overestimated.

5.14 UNCERTAINTIES

Uncertainties may relate to several factors, such as toxicity information or exposure assessments.

5.14.1 Uncertainties Related to Toxicity Information

The RfDs for COPCs are a major source of uncertainty in a BLRA. A chronic RfD is an estimate of the daily exposure unlikely to present an appreciable risk of deleterious effects during a lifetime. Uncertainty factors are applied to selected exposure levels identified in animal or human studies to derive the RfD. To avoid overestimating the RfD, identified exposure levels are divided by these uncertainty factors. An uncertainty factor of 10 is used to account for variations in human sensitivity when using data from valid human studies involving long-term exposure of average, healthy subjects. When extrapolating from observations of toxicity in animals to predicted toxicity in humans, additional uncertainty factors of 10 are applied.

Uncertainties are also present when identifying whether a compound is a likely human carcinogen and at what level of exposure an increased risk of cancer may exist. Uncertainties in quantifying the exposure levels that may result in elevated carcinogenic risk for specific compounds are corrected for by using the 95% UCL of the slope relating exposure to the probability of developing cancer. The actual slope may be greater, but is unlikely to be greater.

The lack of an RfD and a cancer SF for lead will introduce uncertainty into the BLRA if lead is a COPC.

5.14.2 Uncertainties in the Exposure Assessment

Uncertainties in the exposure assessment are introduced in estimating the concentrations to which receptors may be exposed to in the future and in identifying exposed populations. The future land use at the facility and prediction of the future surrounding use add uncertainty to this assessment. The exposure scenarios selected are developed to model the highest reasonable potential exposures to site contaminants. These estimates are unlikely to underestimate future potential risk.

Estimates of exposure frequency and duration are also uncertain. Reasonable levels were selected that are not likely to underestimate the risk associated with site-related activities.

- Uncertainty is present in exposure point concentration estimations.
- All data for metals in groundwater are based on unfiltered groundwater samples.

5.15 RISK CHARACTERIZATION

Based on intake calculations and the identification of complete exposure pathways, an overall site characterization and a risk characterization are completed. This will include the following:

- Written justification for the assumptions used to calculate human dose or intake
- Characterization of carcinogenic risk using EPA-established carcinogenic SFs
- Estimation of carcinogenic risk expressed as the incremental increase in the probability of an individual developing cancer over a lifetime (incremental lifetime cancer risks [ILCRs])
- Summation of ILCRs for individual COPCs across pathways and within receptor exposure scenarios (e.g., on-site worker exposed to groundwater)
- Estimation of potential adverse health effects from exposure to systemic toxicants via comparing an exposure intake to a standard RfD (This ratio is the HQ.)
- Estimation of HQs for each COPC for which toxicity values are available
- Assumption of individual COPC HQ additivity applied to chemicals that induce the same effect on the same target organ

The summation of HQs is judged to form valid upper-bound hazard indices. These summations are considered only when chemicals within the mixture exhibit "dilution type interaction" (Science Advisory Board Review of the OSWER draft RAGS, Human Health Evaluation Manual [HHEM], EPA-SAB-EHC-93-007). These chemicals within a mixture must have interactions that are independent mechanisms; synergistic and/or antagonistic interactions invalidate the summation of HQs.

Consider indirect exposures (e.g., through the food chain) when receptors have been identified as currently present on-site or potentially identified given reuse options. Grazing scenarios are to be considered both as indirect exposures to humans and with the ecological assessment.

Consider exposure pathways related to soil contamination in terms of dermal and inhalation of fugitive dust hazards.

5.16 HEADSPACE MONITORING—VOLATILES

The PID is a quantitative instrument that measures the total concentration of various VOCs in the air. The PID may be used as an approach instrument to monitor for safe approach to the site's hot spots and also for headspace analysis of any samples taken.

When wells are drilled and/or soil borings are taken, the headspace in the borehole is monitored to assure safety to the drill crew. The PID measures in the parts per million range; therefore, sustained deflection of over 5 ppm for 1 min is a good indicator of volatile presence long before most volatile chemicals reach an explosive potential.

PIDs are also used for ongoing monitoring of personnel exposures. If a detection of volatiles occurs, either detector tube or solvent tube sampling may be required to identify the exact volatile chemical constituency.

5.17 O$_2$/CGI

The O$_2$/CGI is an air-monitoring device designed to indicate the level of oxygen present and monitor for a flammable/explosive atmosphere. The CGI registers combustible gas or vapors in terms of their LEL, which is the lowest concentration at which a combustible gas may ignite (or explode) under normal atmospheric conditions.

These instruments are required on all sites where volatiles may be expected to reach LEL levels and for all sampling requiring confined space entry.

5.18 INDUSTRIAL MONITORING—PROCESS SAFETY MANAGEMENT

Sampling at industrial sites to determine chemical risk proceeds to determine employee exposure potential and ultimately chemical risk. The chemical risk assessment scenarios used by the EPA may or may not be applicable or relevant. In cases where exposure can be compared to OSHA PELs and STELs, calculation of risk may not be necessary.

Screening with portable monitors (PIDs and O_2/CGIs) or detector tubes can be used to evaluate the following:

- Exposures to substances as to PELs in relatively dust-free atmospheres
- Intermittent processes using substances that do not have STELs
- Engineering controls
- Work practices
- Isolation of process

A sufficient number of samples must be taken to obtain a representative estimate of exposure. Contaminant concentrations vary seasonally, with weather, with production levels, and in a single location or job class. When determining exposure levels, you may elect to turn off or remove sampling pumps before employees leave a potentially contaminated area (such as when they go to lunch or on a break). If you follow this OSHA-allowed protocol, you MUST document and be able to prove zero exposure during the time interval the monitor was turned off.

5.19 BULK SAMPLES

Bulk samples are often required to assist the industrial hygienist in the proper analysis of field samples at industrial sites. Bulk samples can also be taken and analyzed to support any hazard communication inspections (i.e., Material Safety Data Sheet determinations).

Biological Risk Assessment

Given that many of the indoor air problems (whether on remediation sites or elsewhere) are caused by biological contaminants and their decomposition products, Chapter 6 provides real-world examples of biological monitoring protocols.

All sampling for biologicals must take into account surrounding environmental factors and building usage. Drawing a complete history and in some cases additional types of air sampling for other contaminants are required. In discussing sampling for biologicals we will also discuss the reasons for concern and control mechanisms that can be used.

You must always remember that biologicals, unlike chemical contaminants, have the potential to reproduce and thus grow in numbers. Care must be taken to sample in a consistent fashion in as short a period as possible.

Reproduction of biologicals also calls into question the relative viability of spores and bacterial colonies that are encysted. In cases where amplification is primarily bacterial, these colonies may inhibit spores from developing into vegetative structures. Consequently comparative levels for bacterial counts and mold colony forming units (CFUs) may be required, especially since spores in and of themselves can be problematic. Thus, the absence of visible mold growth may not be indicative of a clean environment.

When molds are amplified to the extent that the building is increasingly hospitable to further mold growth, we may begin to see pathogenic colonies (that would not otherwise be present) taking hold in a building's interior. All of us exhibit great concern when confronted with possible *Stachybotrys atra* (Stachy). Airstream movement does not readily spread Stachy, as the spores become less viable in dry airstream environments. However, in moisture-laden airstreams or within homes with other amplified mold colonies, Stachy may begin to flourish.

In areas where bird or other animal droppings are prevalent, we begin to be concerned about histoplasmosis and coccidiomycosis. *Histoplasma* (Histo) is the more likely disease vector where other molds are flourishing, given that Histo is better able to survive in wetter environments.

Keep in mind that the term *wet* is a relative one. Some of these molds do not need "wet" environments in the traditional sense to grow well; any condensation will do, even that caused by very slight temperature differences.

The old way of thinking that fiberglass will not grow biologicals is also not correct. The fiberglass itself may not be a good food source; however, the fiberglass forms a nice nest and traps other food sources. Fiberglass filters, lined fiberglass ducts, and fiberglass panels inserted for insulation all become less densely packed with age and use. Particulates, especially those associated with any greasy, vapor-laden airstream, stick to the fiberglass and provide a nutrient bed for biological contamination.

Because of the problems with grease or oil in airstreams and biological amplification, care must be taken in using these products. Whenever refrigerant lines bearing mineral oil and freon are serviced, any breakage should be viewed as potentially providing a nutrient "fly paper" for biological contaminants.

So—how much is bad? This is determined in part by aesthetic concerns and in part by health concerns. If you do not want visible mold growth, even small colonies may be too much. Larger colonies, even if no health effects are forthcoming, are certainly unacceptable and over time may even do structural damage. *Aspergillus* can thrive on cellulose, paint, and drywall, leaving unpleasant looking stains as the colonies die. The health questions have many answers depending in part on how sampling is accomplished. With current sampling protocols we become concerned if any part of a building is showing amplified mold growth. We often compare to exterior background levels or to levels in a part of the building shown to be relatively free of mold contamination. In the sense that these biological contaminants may be ever changing in numbers as conditions change, there is no such thing as a static background level. The lack of "hard numbers" is one other reason that the sampling team and microbiologist oversight must include senior level scientists.

For sensitized individuals, the elderly or very young and immune-compromised people, even very sparse mold colonies may cause health problems. Certainly anyone hospitalized for surgery or other invasive medical procedures would also be considered immune compromised during that interval of time. For individuals without these types of concerns, we want to see nonpathogenic mold counts less than 200 CFU/m^3 over established background levels. Higher levels may be acceptable for certain mixes of mold species, and lower levels are required for single species and pathogenic contaminant confidence.

Contact samples should always be less than 200 CFU/strip for areas to be judged "clean." A combination of contact and air-sampling information is required to assess most buildings (Figure 6.1), and these acceptable numbers vary given different biological contaminant mixes and building usage. For example, in a hospital setting, 20 CFU/strip would be too much in the operating room and perfectly acceptable in the visitor's waiting room.

Once the level of contamination is assessed, we can begin to decide how to remedy any negative situations. Steam cleaning without the use of biocides is sometimes the wrong thing to do. Remember that even steam cools, and cool water is just what most molds need

Figure 6.1 Biological contact agar strips. (Biotest Diagnostic Corp.)

to increase their amplification rate. Steam cleaning can be beneficial with adequate drying cycles and in some instances the concurrent use of biocides.

Biocide usage can also be problematic. Chemicals that work in the laboratory may cause aesthetic and even health problems in the real world. Biocides often have limited residual time and may not even be tested against the particular biological contaminant mix of concern.

Residual time for any chemical mix has many unresolved questions; sometimes the chemicals' residual time in your particular circumstances is not even known. In other cases residual time may make the chemical unattractive because the toxic properties of the chemical remain and can cause contamination problems in and of themselves. We must always remember that the basic cellular structure of these biological contaminants and ourselves is the same, so chemicals that harm these contaminants may also harm us.

Recently, biocides with FDA and EPA approval have been developed and can be used in areas where biocides were formerly unacceptable. The decision in selecting the appropriate biocide that will not harm humans, animal occupants, or damage structural materials can be a difficult one. Decontamination and rehabitation methodologies must be part of a coordinated remedial design effort.

One of the more common replies to all of this information is—why now? The answer is twofold; first, we probably always had these concerns once we lived for any length of time indoors; second, we have increasingly closed our buildings and relied on forced air ventilation systems. Both of these answers are also applicable to closed cab modes of transportation—airplanes, automobiles, rail cars, and ships.

In the past we thought that endpoint filtration of airstreams was sufficient to render delivered air relatively pure. We have now learned that filtration only works for a time, and excessive biological amplification can be transmitted through most current HVAC systems once established in ductwork or plenums.

If you suspect biological contamination, see visible mold growth, have personnel with repetitive mycosial infections, or have indoor air quality (IAQ) problems that have remained undiagnosed, you need to consult a team of professionals to find answers to these problems. In these cases not only is the mold growth itself problematic, but also we have to worry about the chemicals formed as colonies die. Dieback causes the chemicals formed during decomposition to be spread throughout building, and these are the same VOCs we worry about from chemical spills or misuse. The following sections speak directly to hazards associated with mold and fungi.

6.1 FUNGI, MOLDS, AND RISK

When inhaled, microscopic fungal spores or fragments of fungi may cause allergic rhinitis. Because they are so small, mold spores may evade the protective mechanisms of the nose and upper respiratory tract to reach the lungs and bring on asthma symptoms. The buildup of mucus, wheezing, and difficulty in breathing are the result. Less frequently, exposure to spores or fragments may lead to a lung disease known as hypersensitivity pneumonitis.

Molds are present in our exterior environments, and, hopefully, to a lesser extent in our interior environments. People allergic to molds may have allergic symptoms from spring to late fall. The mold season often peaks from July to late summer. Unlike pollens, molds may persist after the first killing frost. Some can grow at subfreezing temperatures, but most become dormant. Snow cover lowers the outdoor mold count drastically, but does not kill molds. After the spring thaw, molds thrive on the vegetation that has been killed by the winter cold. In the warmest areas of the world, however, molds thrive year-round and can cause perennial allergic problems. Molds growing indoors can cause perennial allergic rhinitis even in the coldest climates.

If indoor areas show signs of amplification identified by visual assessment, air sampling, and contact/liquid sampling, amplification must be suspected. Amplification is the process whereby biological organisms continue to increase over time. If this increase is not controlled, sufficient mold spores and vegetative structures may be present to create indoor air problems.

Hot spots of mold growth in the home include damp basements and closets, bathrooms (especially shower stalls), places where fresh food is stored, refrigerator drip trays, house plants, air conditioners, humidifiers, garbage pails, mattresses, upholstered furniture, and old foam rubber pillows.

6.1.1 What Is the Difference between Molds, Fungi, and Yeasts?

Molds and yeasts are two groups of plants in the fungus family. Yeasts are single cells that divide to form clusters. Molds consist of many cells that grow as branching threads called hyphae. The seeds or reproductive particles of fungi are called spores. They differ in size, shape, and color among species. Each spore that germinates can give rise to new mold growth, which in turn can produce millions of spores.

6.1.2 How Would I Become Exposed to Fungi That Would Create a Health Effect?

The route of exposure may be inhalation or ingestion accompanied by inhalation. When inhaled, microscopic fungal spores or fragments of fungi may cause health problems. Because they are so small, mold spores may evade the protective mechanisms of the nose and upper respiratory tract to reach the lungs and bring on asthma symptoms. The buildup of mucus, wheezing, and difficulty in breathing are the result. Less frequently, exposure to spores or fragments may lead to a lung disease known as hypersensitivity pneumonitis.

6.1.3 What Types of Molds Are Commonly Found Indoors?

In general, *Alternaria* and *Cladosporium* (*Hormodendrum*) are the molds most commonly found both indoors and outdoors throughout the U.S. *Aspergillus, Penicillium, Helminthosporium, Epicoccum, Fusarium, Mucor, Rhizopus,* and *Aureobasidium* (*Pullularia*) are also common.

6.1.4 Are Mold Counts Helpful?

Similar to pollen counts, mold counts may suggest the types and relative quantities of mold present at a certain time and place. For several reasons, however, these counts probably cannot be used as a constant guide for daily activities. One reason is that the number and types of spores actually present in the mold count may have changed considerably in 24 h because weather and spore dispersal are directly related. Many of the common allergenic molds are of the dry spore type—they release their spores during dry, windy weather.

Other molds need high humidity, fog, or dew to release their spores. Although rain washes many larger spores out of the air, it also releases some smaller spores into the air.

6.1.5 What Can Happen with Mold-Caused Health Disorders?

Fungi or microorganisms related to them may cause other health problems similar to an allergy. Fungi may lodge in the airways or a distant part of the lung and grow until they form a compact sphere known as a "fungus ball." In people with lung damage or serious underlying illnesses, *Aspergillus* may grasp the opportunity to invade and actually infect the lungs or the whole body. In some individuals exposure to these fungi can also lead to asthma or to an illness known as "allergic bronchopulmonary aspergillosis." This latter condition, which occurs occasionally in people with asthma, is characterized by wheezing, low-grade fever, and coughing up of brown-flecked masses or mucous plugs. Skin testing, blood tests, X-rays, and examination of the sputum for fungi can help establish the diagnosis. The occurrence of allergic aspergillosis suggests that other fungi might cause similar respiratory conditions.

Inhalation of spores from fungus-like bacteria, called actinomycetes, and from molds can cause a lung disease called hypersensitivity pneumonitis. This condition is often associated with specific occupations. Hypersensitivity pneumonitis develops in people who live or work where an air-conditioning or a humidifying unit is contaminated with and emits these spores. The symptoms of hypersensitivity pneumonitis may resemble those of a bacterial or viral infection such as the flu. If hypersensitivity pneumonitis is allowed to progress, it can lead to serious heart and lung problems.

6.2 BIOLOGICAL AGENTS AND FUNGI TYPES

A host of fungi are commonly found in ventilation systems and indoor environments. The main hazardous species belong to the following genera: *Absidia, Alternaria, Aspergillus, Fusarium, Cladosporium, Cryptostroma, Mucor, Penicillium,* and *Stachybotrys.* Various strains of these genera of molds have been implicated in being causative agents in asthma, hypersensitivity pneumonitis, and pulmonary mycosis.

Fungi commonly found in ventilation systems and indoor environments include *Absidia, Acremonium, Alternaria, Aspergillus, Aureobasidium, Botrytis, Cephalosporium, Chrysosporium, Cladosporium, Epicoccum, Fusarium, Helminthosporium, Mucor, Nigrospora, Penicillium, Phoma, Pithomyces, Rhinocladiella, Rhizopus, Scopulariopsis, Stachybotrys, Streptomyces, Stysanus, Ulocladium,* Yeast, and *Zygosporium.* Eleven types of fungi are typically found in homes: *Aspergillus, Cladosporium, Chrysosporium, Epicoccum, Fonsecaea, Penicillium, Stachybotrys, and Trichoderma.*

6.2.1 *Alternaria*

A number of very similar, related species are usually grouped together as *Alternaria.* The spores of *Alternaria* are multicelled and developed in chains, head-to-toe, from which their name derives. Spores are multiseptate, both transverse and longitudinally. They vary in width and length according to species, usually 8–75 μm long; some species such as *A. longissima* are up to 0.5 mm long. *Alternaria,* which is both ubiquitous and abundant, is both saprophytic and parasitic on plant material and is found on rotting vegetation as well as in damp indoor areas, such as bathrooms. Some species of *Alternaria* are the imperfect, asexual, anamorph spores of the ascomycete *Pleospora.*

6.2.2 *Aureobasidium*

Aureobasidium is common in both outdoor and indoor air, bathroom walls, and shower curtains. *Aureobasidium* causes mildew and has been isolated in flooded areas of buildings, as well as from soils, plants, and other substrates. *Aureobasidium* has been associated with hypersensitivity pneumonitis in some individuals.

6.2.3 *Cladosporium*

Cladosporium, composed of over 500 species, is found in outdoor as well as indoor air. *Cladosporium* has been isolated from fuels, wood, plant tissues, straw, face cream, air, soil, foods, paint, and textiles. *Cladosporium* spores are often found in higher concentrations in the air than any other fungal spore type.

Cladosporium bears copious numbers of spores on branched conidiophores. The spores usually have distinctive "scars" at both ends where they are joined both to the spore at one end and to the conidiophore at the other. Although often identified as single-celled spores, spores are frequently seen with a single transverse septum or several transverse septa. Their length ranges from 4 to 20 μm.

Cladosporium (*Hormodendrum*) is the most commonly identified outdoor fungus and is a common indoor air allergen. Indoors *Cladosporium* may be different from the species identified outdoors. *Cladosporium* is commonly found on the surface of fiberglass duct liners in the interior of supply ducts. *Cladosporium* can cause mycosis and is a common cause of extrinsic asthma (immediate-type hypersensitivity: type I). Acute symptoms include edema and bronchiospasms; chronic cases may develop pulmonary emphysema.

6.2.4 *Rhodotorula*

Rhodotorula is a commonly isolated yeast that is frequently isolated from humidifiers and soil. *Rhodotorula* may be allergenic to susceptible individuals when present in sufficient concentrations.

6.2.5 *Stemphylium*

Stemphylium is a saprophytic fungus (grows on nonliving organic material) commonly found on cellulosic materials (that is, of plant origin, including livestock feed, cotton cloth, ceiling tiles, paper). *Stemphylium* is an example of a diurnal sporulator. An alternating light and dark cycle is required for spore development. This fungus requires ultraviolet light for the production of conidiophores; however, the second developmental phase, when the conidia are produced, requires a dark period. *Stemphylium* also requires wet conditions for growth. Stemphylium spores range from 23 to 75 μm in length.

6.2.6 Sterile Fungi

Sterile fungi are common to both outdoor and indoor air. These fungi produce vegetative growth, but yield no spores for identification. Their presence will increase CFU/l. Derived from ascospores or basidiospores, the spores of which are likely to be allergenic, these fungi should be considered allergenic.

6.2.7 Yeast

Various yeasts are commonly identified on air samples. Yeasts are not known to be allergenic, but they may cause problems if a person has had previous exposure and developed hypersensitivities. Yeasts may be allergenic to susceptible individuals when present in sufficient concentrations. Yeast grows when moisture, food, and just the right temperatures are available.

6.3 *ASPERGILLUS*

Aspergillus and *Penicillium* are molds prevalent in soils. These molds can cause asthma-like symptoms or other lung irritation in humans and deterioration in buildings and other materials. When conditions within buildings cause the buildup of moisture on surfaces and temperatures are right, *Aspergillus* grows well and is evidenced by a black deposit.

Aspergillus is a type of mold called Ascomycota or sac fungi. Sac fungi have sexual spores that are produced in an ascus or saclike structure. Their asexual spores, called conidiospores (from the word *conidia*, which means "dust"), are produced in long chains from a conidiophore. The characteristic arrangement of the conidiospores is used to identify the different molds. *Penicillium* is another mold that is also called *Ascomycetes*.

6.3.1 What Color Are These Molds?

Aspergillus is black, and *Penicillium* is white. Also, *Aspergillus* is not the black mold on bread. That mold is *Rhizopus nigricans*. The difference is evident in the differing structures for black asexual spores (sporangiospores).

6.3.2 How Is *Aspergillus* Spread?

Aspergillus spores are carried in the wind and through ventilation airstreams in homes. The asexual spores freely detach from the conidiophore chain and, with the slightest disturbance, float in the air like dust. The easiest way to get *Aspergillus* started in the home is to bring the spores in on shoes and deposit the spores on carpet fibers.

6.3.3 How Does *Aspergillus* Grow/Amplify?

When the spores are placed on wet surfaces, the spores grow hyphae. The hyphae grow, form a mass, and are soon visible to the naked eye. The vegetative mycelium process foods, and reproductive mycelium create more spores. At this time the mold/fungi appears as a black fuzzy mass. (*Amplification* is the process whereby *Aspergillus* or other biological organisms continue to increase in number over time.)

6.3.4 What Conditions Help *Aspergillus* Grow/Amplify?

Fungi generally grow better with an acidic pH. The growth is usually on the surface rather than embedded within a substrate (under the surface).

Fungi are able to grow on surfaces with a low moisture content, in contrast to the moisture required for bacterial growth. Therefore, even a slight difference in temperature and surface moisture facilitates the growth of fungi.

Fungi are capable of using complex carbohydrates, such as lignin (wood). Thus, with a little moisture, fungi can easily grow on wood or other complex organic materials. These adaptations allow fungi to grow readily on painted walls and shoe leather.

6.3.5 Can Mold/Fungi Make You Sick?

Fungal diseases are called mycoses, which are chronic, long-lasting infections. Aspergillosis is an opportunistic infection that can become pathogenic (disease-causing) in a weakened individual host. The inhalation of spores is a possible mode of entry into the body as spore size ranges from 2 to 10 μm.

6.3.6 What Are the Symptoms of Aspergillosis?

The incubation period varies with different individuals. People with other weakening medical problems or general ill health are most susceptible. *Aspergillus niger* (*A. niger*) produces mycotoxins that can induce asthma-like symptoms. In situations when *A. niger* was found growing with *Penicillium* sp., massive inhalation of spores has been documented as causing an acute, diffuse, self-limiting pneumonitis (lung irritation). Healthy individuals can exhibit otitis externa (inflammation of the outer ear canal) as a result of *Aspergillus* growth.

6.3.7 Does *Aspergillus* Cause Deterioration of Materials?

Members of the *Aspergillus* genus are known as biodeteriogens (organisms that cause deterioration of materials). *A. niger* causes damage, discoloration, and softening of the surfaces of woods, even in the presence of wood preservatives. *A. niger* also causes damage to cellulose materials, hides, and cotton fibers. *A. niger* can also attack plastics and polymers (i.e., cellulose nitrate, polyvinyl acetate, polyester type polyurethanes).

6.3.8 What Happens If *Aspergillus* Colonies Grow inside Construction Layers?

In cases of extensive growth, colonies will grow into wood, plaster, and/or drywall, causing a soft bulging area. This area lacks structural integrity and is subject to early deterioration.

6.3.9 How Is *Aspergillus* Identified?

Soy agar will grow *Aspergillus* and a wide range of other microbiologicals. Thus, Tryptic Soy Agar or Potato Dextrose Agar is the original screening tool used to determine the presence of biologicals. Once biological contamination has been established, selective media can be used to grow suspect organisms for identification. Using a special type of protein gelatin (called Rose Bengal Agar) that has been made with special nutrients, *Aspergillus* cultures can be selectively and quickly grown.

6.3.10 How Are Levels of *Aspergillus* Communicated?

Aspergillus is reported in terms of colony forming units per cubic meter. The presence of any one fungi in excess of 200 CFU/m^3 is indicative of an indoor source of fungal amplification. The presence of any colony forming units per cubic meter is indicative of transmission of fungal spores from surface to surface and/or from exterior to interior locations.

6.3.11 Why Do *Aspergillus* Colonies Look Black?

Aspergillus is black or brown-black. Also, active biological contamination creates a surface to which dusts and other debris "stick." If biological contamination is extensive and characterized by amplification and "kill" cycle condition, the fungi/molds will decay and produce toxins. These toxins can be identified with *Aspergillus* contamination as a black stain or tarlike liquid residue.

6.3.12 What Will Biotesting of the Air Show?

Biotesting using a BIOTEST air monitor will reveal whether colony forming units are found in the air. Biotesting by surface culturing on agar reveals the presence of biologicals on surfaces and in waters.

6.3.13 What Can Be Done to Prevent *Aspergillus* Growth?

Keep the air dry, provide filtered replacement air, and have sufficient air exchanges. Prevent accumulation of standing water or leaks.

6.4 *PENICILLIUM*

Penicillium is a very large group of fungi valued as a producer of antibiotics. *Penicillium* is commonly found in the soil; in the air; on living vegetation, seeds, grains, and animals; and on wet insulation. *Penicillium* has been associated with hypersensitivity pneumonitis in some individuals when it is present in high concentrations.

Penicillium is a source of antibiotic lines that have aided humanity. However, not all species of *Penicillium* are helpful. Some can cause allergic reactions and other adverse health effects when dispersed through indoor air. Currently, more and more is being learned about the effects of *Penicillium* and other microbiologicals in indoor air. This section represents a starting discussion of the risks associated with the growth of *Penicillium* within indoor air environments.

Penicillium is a fungus that grows when moisture, food, and just the right temperatures are available. *Penicillium*'s spherical spores are produced in long, unbranched chains of each conidiophore. These usually fragment into individual spores, although chains of spores are seen periodically on slides. Although some species of *Penicillium* appear to reproduce solely by asexual means, some species of *Penicillium* are the anamorph (asexual) stage of the ascomycete genus *Talaromyces*.

6.4.1 What Do Samples Look Like?

When samples are freshly prepared from culture, the spores are pale green, although this fades with age. Their size ranges from 3 to 5 μm. When using visual methods of identification, *Aspergillus* and *Penicillium* cannot be differentiated because the spores are so similar that they are grouped together into the *Aspergillus/Penicillium* group. Spores from this group are found almost all year-round.

6.4.2 What Species of *Penicillium* Are Used to Produce Antibiotics?

Penicillin, as produced by Alexander Fleming in 1929, was a product of *Penicillium notatum*. Since that time, other species of *Penicillium* have been used to form other antibiotics. As an example, Griseofulvin is an antifungal antibiotic formed from a species of *Penicillium*.

6.4.3 What Other Fungi Grow Where *Penicillium* Grows?

Aspergillus, Penicillium, Verticillium, Alternaria, and *Fusarium* are all found in the order Moniliales and have similar morphology. Thus, where *Aspergillus* is found, one may expect to find *Penicillium* and vice versa. The key here is the relative presence of moisture that may accelerate the growth of one particular fungus rather than another.

6.4.4 If *Penicillium* Grows Everywhere, What Is the Concern?

The concern is that, in most cases, we do not want *Penicillium* growing inside us. This warning is especially true if an individual is immune compromised.

People sensitized to *Penicillium*, the very young, the aging population, and people with certain illnesses, could be considered immune compromised. These individuals may react more strongly (and often more negatively) to some *Penicillium* species entering their bodies.

6.4.5 How Does *Penicillium* Enter the Body?

The route of entry into the body is unknown. However, the respiratory route is used by many other fungi with abundant conidia. *Penicillium* may have abundant conidia; thus, the respiratory route of entry is expected. Skin trauma has been associated with local infection, but not with systemic disease. Infection via the digestive route is unusual for filamentous fungi.

6.4.6 Are There Particular Species of *Penicillium* about Which I Should Be Concerned?

Within current medical literature, the primary concern is with *Penicillium marneffei* (*P. marneffei*). This species has two life formations and is the only *Penicillium* species that is termed *dimorphic*. The prevalence of one form over another is dependent on temperature. At 37°C the fungus grows as yeasts forming white-to-tan, soft, or convoluted colonies. Microscopically, the yeasts are spherical or oval and divide by fission rather than budding.

At 25°C the fungus produces a fast-growing, grayish floccose colony. Microscopic examination reveals septate branching hyphae with lateral and terminal conidiophores that produce unbranched, broomlike chains of oval conidia.

Inside the body *P. marneffei* first proliferates in the reticuloendothelial system and then is disseminated. The lungs and liver are usually the most severely involved organs. Other commonly involved organs include skin, bone marrow, intestine, spleen, kidney, lymph nodes, and tonsils.

The reticuloendothelial system is made up of special cells called phagocytes located throughout the body; they can be found in the liver, spleen, bone marrow, brain, spinal cord, and lungs. When functioning correctly, phagocytes destroy disease-causing organisms by ingesting the organisms. An example of these cells are histiocytes. Histiocytes try to ingest and kill *P. marneffei*. Unfortunately when the *P. marneffei* do not die, the histiocytes carry them throughout the body.

6.5 FUNGI AND DISEASE

The main hazardous species belong to the following genera: *Absidia, Alternaria, Aspergillus, Fusarium, Cladosporium, Cryptostroma, Mucor, Penicillium,* and *Stachybotrys.* Various strains of these genera of molds have been implicated in being causative agents in asthma, hypersensitivity pneumonitis, and pulmonary mycosis. Fungi commonly found in ventilation systems and indoor environments include *Absidia, Acremonium, Alternaria, Aspergillus, Aureobasidium, Botrytis, Cephalosporium, Chrysosporium, Cladosporium, Epicoccum, Fusarium, Helminthosporium, Mucor, Nigrospora, Penicillium, Phoma, Pithomyces, Rhinocladiella, Rhizopus, Scopulariopsis, Stachybotrys, Streptomyces, Stysanus, Ulocladium,* Yeast, and *Zygosporium.*

6.5.1 *Blastomyces dermatitidis*

Local infections have occurred following accidental parenteral inoculation with infected tissues or cultures containing yeast forms of *B. dermatitidis.* Parenteral (subcutaneous) inoculation of these materials may cause local granulomas.

Pulmonary infections have occurred following the presumed inhalation of conidia; two individuals developed pneumonia and one had an osteolytic lesion from which *B. dermatitidis* was cultured. Presumably, pulmonary infections are associated only with sporulating mold forms (conidia).

6.5.2 *Coccidioides immitis*

Clinical disease may occur in 90% of an exposed indoor population. Infections acquired in nature are asymptomatic in 50% of these outdoor cases. Because of their size (2–5 nm), the arthroconidia are conducive to ready dispersal in air and retention in deep pulmonary spaces. The much larger size of the spherule (30–60 nm) considerably reduces the effectiveness of this form of the fungus as an airborne pathogen. Spherules of the fungus may be present in clinical specimens and animal tissues, and infectious arthroconidia may be present in mold cultures and soil samples. Inhalation of arthroconidia from soil samples, mold cultures, or following transformation from the spherule form in clinical materials is the primary hazard. Accidental percutaneous inoculation of the spherule form may result in local granuloma formation.

6.5.3 *Histoplasma capsulatum*

Pulmonary infections have resulted from handling mold from cultures. Collecting and processing soil samples from endemic areas have caused pulmonary infections in laboratory workers. Encapsulated spores are resistant to drying and may remain viable for long periods. The small size of the infective conidia (less than 5 μm) is conducive to airborne dispersal and intrapulmonary retention. The infective stage of this dimorphic fungus (conidia) is present in sporulating mold from cultures and in soil from endemic areas. The yeast form is in tissues or fluids from infected animals and may produce local infection following parenteral inoculation.

6.5.4 *Sporothrix schenckii*

Sporothrix schenckii has caused a substantial number of local skin or eye infections in laboratory personnel. Most cases have been associated with accidents and have involved splashing culture material into the eye, scratching or injecting infected material into the skin, or being bitten by an experimentally infected animal. Skin infections have also resulted from handling cultures or from the necropsy of animals. No pulmonary infections have been reported to result from laboratory exposure, although naturally occurring lung disease is thought to result from inhalation.

6.5.5 Pathogenic Members of the Genera *Epidermophyton*, *Microsporum*, and *Trichophyton*

Skin, hair, and nail infections by these dermatophytid molds are among the most prevalent of human infections. Agents are present in the skin, hair, and nails of human and animal hosts. Contact with infected animals with inapparent or apparent infections is the primary hazard. Cultures and clinical materials are not an important source of human infection.

6.5.6 Miscellaneous Molds

Several molds have caused serious infection in immunocompetent hosts following presumed inhalation or accidental subcutaneous inoculation from environmental sources. These agents are *Cladosporium* (*Xylohypha*) *trichoides*, *Cladosporium bantianum*, *Penicillium marnefii*, *Exophiala* (*Wangiella*) *dermatitidis*, *Fonsecaea pedrosoi*, and *Dactylaria gallopava* (*Ochroconis gallopavum*). The gravity of naturally acquired illness is sufficient to merit special precautions. Inhalation of conidia from sporulating mold cultures or accidental injection into the skin is a risk.

6.5.7 *Fusarium*

The corn fungus *Fusarium moniliforme* produces fusaric acid that behaves like a weak animal toxin, but combined with other mold toxins, it exaggerates the effects of the other toxins. Scientists consider that this may be the important role for fusaric acid. All isolates of the *Fusarium*-type molds produce this toxin, suggesting that this compound is probably more prevalent in the environment than was initially considered. These results indicate that analyses and toxicity studies should also include this toxin along with other suspect toxins under field conditions. Fumonisins might be teratogenic to humans.

6.6 FUNGI CONTROL

Call in professional help! If you are unsure of the biological condition of your facility or have ongoing unidentified indoor air problems, assume you have a biological emergency. The standing rules are as follows: Bleach what you can bleach. Use biocides with caution. Throw out what you can throw out. If you are unsure about any of these protocols, get help!

6.6.1 Ubiquitous Fungi

Fungi are ubiquitous in the environment, particularly in soil, and many are also part of the normal gastrointestinal and skin flora in humans and animals. In some areas of the U.S., certain types of fungi are endemic and occur naturally in the soil. These soil fungi include *Histoplasma capsulatum*, found in some midwestern states, and *Coccidioides immitis*, which is found in the southwestern U.S. and parts of Central and South America. If the soil habitat of these fungi is disturbed by activities such as construction or natural disasters, the fungal spores become airborne; when they are inhaled, they can cause infection.

6.6.2 Infection

Fungal infections can cause a variety of symptoms. Some types of fungi can infect persons with normal immune systems. Examples are the airborne spores of *Blastomyces dermatidis*, *Coccidioides immitis*, or *Histoplasma capsulatum*, which cause respiratory symptoms ranging from mild illness to pneumonia to severe disseminated disease. Other pathogenic molds, called dermatophytes, cause ringworm infections of the skin, hair, and nails, such as athlete's foot, jock itch, and scalp ringworm. Unlike most fungi these can be transmitted from person to person.

Mushrooms are also fungi, and some can cause life-threatening food poisoning. The fungi that are responsible for the recent increase in mycotic infections are those causing opportunistic infections. These organisms include *Candida* species, *Cryptococcus neoformans*, *Aspergillus* species, *Fusarium* species, *Coccidioides immitis*, and *Histoplasma capsulatum*. Persons at high risk for opportunistic fungal infections are those with HIV infection or AIDS, those who have undergone bone marrow or organ transplants, those receiving chemotherapy for cancer, and others who have had debilitating illness, severe injuries, prolonged hospitalization, or long-term treatment with corticosteroid or antibacterial drugs.

6.6.3 Immediate Worker Protection

Whenever possible, use remote methods for cleanup. If you must enter a biological environment, consult a certified industrial hygienist and microbiologist with experience in biological environments. At a minimum, when entering an area where any invasive activities will occur, use HEPA filters for pulmonary protection and wear dermal protection for hands. All material worn or used must be either decontaminated or properly disposed.

6.6.4 Decontamination

Decontamination may consist of washing with chlorinated or other oxidizing chemicals (i.e., bleach or oxidizing, color-safe bleach; ozone). Biocides may also be used; however, make sure that the biocides are proven effective for the particular biologicals present. All of these chemicals have in and of themselves some risk to workers. For porous surfaces, including fiberglass liners inside ducts, encapsulation of the porous surface may be required.

6.6.5 Fungi and VOCs

As molds and fungi grow, they give off emissions of metabolic gases that contain VOCs. Some of the volatile compounds that we have found are primary solvents. Many of these emitted chemicals are identical to those originating from solvent-based building materials and cleaning supplies, including hexane, methylene chloride, benzene, and acetone.

6.6.6 Controlling Fungi

Moisture, heat, and dirt or dusts are the ingredients needed to grow fungi. Remove one or more, and you'll likely have a healthier building.

Furnace filters—often used in homes, schools, and small office buildings—don't catch and trap all the microbes and the dust the microbes feed upon. Consequently, the filters transmit the microbes and catch large particulates for a time and release a certain percentage of the smaller particles (including spores). Then as the filters become dirtier, the filter material itself begins to catch more of the microbes, provides a growth location, and transfers some of the microbial contamination into the airstream.

Another hot spot for microbial growth is the humidifier assembly on furnaces. Typical humidifier reservoirs are pools of standing, stagnant water throughout much of the year that allow mold to grow and infiltrate the ducts.

In many buildings, fiberglass-lined ductwork (often used for noise or thermal control) is used in lofts because of continual airflow. These lofted spaces collect dirt and become "microbial nests." The microbes grow and multiply and then are blown all over the building to infest other areas.

6.7 ABATEMENT

Should removal of contaminated items be the only option, workers and building occupants must be protected during the abatement. Control of airstreams in and out of the contaminated area is a requirement to limit contamination in other areas of building.

These biologicals grow and reproduce; therefore concentrations of biologicals are not equivalent to such things as chemical concentrations. A dilute concentration of biologicals in a good growth environment will result in a concentrated level of contamination over time.

All abated buildings must be sampled and certified as suitable for reentry prior to normal building usage. This certification states that the building's biological contamination has been diminished through abatement activities and is now in equilibrium with the ambient exterior conditions or interior baseline conditions previously agreed upon.

Certification does not state that the building cannot be recontaminated in the future. Always ask for recommendations for preventing future recontamination of your facility. Any porous materials that have been contaminated, removed from the facility, and returned must be decontaminated; facility users must be advised that recontamination may be inevitable.

CHAPTER **7**

Indoor Air Quality and Environments

This chapter evaluates the operations and maintenance decisions that must be made for air monitoring appropriate for testing ventilation adequacy. It includes a discussion of current air-monitoring instrumentation and methodology.

7.1 VENTILATION DESIGN GUIDE

Mechanical designs should be economical, maintainable, and energy efficient, with full consideration given to the functional requirements and planned life of the facility. Mechanical design should also consider life-cycle operability, maintenance, and repair of the facility and real property–installed equipment, components, and systems. Ease of access to components and systems in accordance with industry standards and safe working practices is a design requirement.

The best way to prevent IAQ problems is to have appropriate and effective engineering controls in place to maintain the indoor air quality. The following is an example of design criteria guidance that should be discussed throughout the design phase. Various boxes throughout the chapter illustrate the real-world concerns from which this guidance was derived.

7.2 EXAMPLE DESIGN CONDITIONS GUIDANCE

The following conditions should be used and will need to be investigated in designing the mechanical systems:

- *Site Elevation:* Equipment design elevation is _____ {insert} feet (meters) above sea level. Appropriate corrections should be made when calculating the capacity of all mechanical equipment installed at this elevation.
- *Latitude:* _____ {insert} Deg N
- *Heating Degree Days:* _____ {insert} annual
- *Cooling Degree Days:* _____ {insert} annual

7.2.1 Outside Design Conditions

Winter:

_____ {insert} °F (°C) for outside makeup air and infiltration loads

_____ {insert} °F (°C) for air transmission loads

Summer:

_____ {insert} °F (°C) dry bulb

_____ {insert} °F (°C) maximum condensation wet bulb

7.2.2 Inside Design Conditions

Winter:

_____ {insert} °F (°C) for occupied administration areas

_____ {insert} °F (°C) for mechanical/electrical areas

Summer:

_____ {insert} °F (°C) for occupied administration areas

_____ {insert} °F (°C) for mechanical/electrical areas

7.3 MECHANICAL ROOM LAYOUT REQUIREMENTS

Mechanical equipment room layout should have ample floor space to accommodate routine maintenance of equipment and adequate headroom to accommodate specified equipment. Ample space should be provided around equipment to allow unobstructed access for servicing and routine maintenance. This space allotment should include ample areas for service and/or replacement of coils, tubes, motors, and other equipment.

Provisions for installation and future replacement of equipment should be coordinated with the architectural design. The arrangement and selection of mechanical equipment should not interfere with complete removal of the largest piece of equipment without dismantling adjacent systems or structures. Doors should be located to facilitate such service.

7.4 ELECTRICAL EQUIPMENT/PANEL COORDINATION

Arrangement of all mechanical equipment and piping should be coordinated with electrical work to provide dedicated space for panels, conduit, and switches. Clearance required by the NEC above and in front of electrical panels and devices should be provided. Mechanical equipment (pipes, ducts, etc.) should not be installed within space that is dedicated to electrical switchboards and panel boards (see NFPA 70 Article 384-4). When electrical equipment is located in a mechanical equipment room, dedicated electrical space including a proper safety envelope must be available.

7.5 GENERAL PIPING REQUIREMENTS

As applicable, the following should be provided for all piping systems:

- All pumps, regardless of design service, should be nonoverloading during operation so the pump can operate at any point on its characteristic pump curve.
- Air vents should be installed on all high points in piping systems.

Air vent location is critical to air actually being vented versus just moving to the next lower air pressure area of the piping. Air and the odors associated with volatile components in the air accumulate in pipes when there is inadequate venting. Ultimately this air is then available to the building proper if there is an "escape route" from the piping. Common escape routes are dry floor drains, through toilet waters, across sink traps and into sink drain-head spaces, and any breaks in piping.

- Valves
  —Vent and drain valves with hose-end connections should be provided on all mechanical systems.
  —Drain valves should be installed at low points and for equipment that must be dismantled for routine servicing.
  —Isolation valves, balancing valves, flow measuring devices, and pressure/temperature test plugs should be provided at all heating and/or cooling terminal units.
  —Bypass piping with isolation valves should be provided around all nonredundant control and system regulating valves.
- Pipe taps, suitable for use with either a 0.125 in. (3.2 mm) outside diameter (OD) temperature or pressure probe, should be located at each pressure gauge.
- Coils
  —All coils should be provided with valved drain and air vent connections.
  —On air-handling units with multiple coils, isolation valves should be installed on the supply piping and a balancing valve on the return piping of each coil.
  —A thermometer should be installed on the supply piping of each coil.
  —Temperature/pressure taps should be provided on the supply and return piping of each coil.
- Strainers should be provided with a valved blowdown connection and piped to a floor drain.
- All underground metallic lines, fittings, and valves, except for cast iron soil and storm drain piping systems, should be cathodically protected.
- All exterior, underground nonmetallic piping should be buried with pipe detection tape.

These design criteria ensure that system components can be located and isolated for maintenance. Areas where piping will be breached after isolation should be identified because in these areas exposure to workers and the environment from pipe contents is most likely during maintenance events. Identification of these areas should then be keyed to general building ventilation systems, location of PPE, and provisions for emergency exiting of the building proper.

7.6 ROOF-MOUNTED EQUIPMENT

Except for intake or relief penthouses, no mechanical equipment should be located on the roof of the facility.

7.7 VIBRATION ISOLATION/EQUIPMENT PADS

Provide vibration-isolation devices on all floor-mounted and suspended mechanical equipment that could transmit noise and vibration to occupied areas. All floor-mounted mechanical equipment should be provided with 6-in. (152-mm) housekeeping pads.

Vibration isolation is also important to prevent the transmission of vibrations to nearby equipment, piping, and control systems. The transmission of vibrations is an issue; sustained vibration of equipment may "shake loose" equipment components.

7.8 INSTRUMENTATION

Sufficient instrumentation must be provided to aid maintenance personnel in balancing and/or troubleshooting mechanical systems. During design the following systems should be assessed for instrumentation requirements:

- Media at each change in temperature point and at all mixing points in chilled water and air-handling systems
- Discharges of air handlers
- Chilled water-blending stations
- Chilled water zone return mains

Pressure gauges, thermometers, flow indicators, and sight glasses should be easily read from the adjacent floor. The following design elements should be addressed:

- Isolation valves on each pressure gauge
- Thermometers with separable socket thermo-wells

The removal, repair, or cleaning of flow-measuring devices should be possible without having to shut down the entire system. In order to accomplish integral system isolation, the following installations should be considered:

- A portable meter, with appropriate range, for each type of flow-measuring device installed
- Separate pressure gauges on both the suction end and the discharge end of pumps

The simple fact is that the easier a system is to maintain, the more likely that maintenance schedules will be followed and that prescribed maintenance will be effective.

7.9 REDUNDANCY

Spare parts that are difficult to obtain or are manufacturer unique, and any special service tools, should be obtained and stored prior to system startup.

7.10 EXTERIOR HEAT DISTRIBUTION SYSTEM

The heat distribution system for the structure extends from and includes the point of connection at the existing system to the service entrance.

7.10.1 Determination of Existing Heat Distribution Systems

Generally, any new distribution systems will have to connect to existing distribution systems at the installation. The first step for the designer is to determine what media are available at the installation. The media distributed in these systems are as follows:

- High-temperature hot water (HTHW) (201–450°F [94–232°C])
- Low-temperature water (LTW) (150–200°F [66–93°C])
- Low-pressure steam (LPS) (up to 15 psig [103 kPa])
- High-pressure steam (HPS) (over 15 psig [103 kPa])
- Condensate return (up to 200°F [93°C])

7.10.2 Selection of Heat Distribution Systems

After the medium type has been determined, the heat distribution system type must then be selected. There are four basic types of distribution systems that can be used:

1. Above ground (AG) (high and low profile)
2. Concrete shallow trench (CST)
3. Buried conduit (BC) (preapproved type)
4. Buried conduit (BC) (not preapproved type)

7.10.2.1 AG Systems

AG systems are the least expensive and lowest cost (for labor) maintenance systems available. However, aesthetic reasons may prevent the use of AG systems. These systems are a good application in industrial areas where the entire piping systems are aboveground. The AG system design should include the following:

- Detailed piping layouts
- Pipe support design (low- or high-profile type)
- Piping insulation selection
- Jacketing selection (to protect against moisture)
- Transition details to buried systems
- Vent, drain, and trap designs

7.10.2.2 CST Systems

CST systems are the preferred buried system. These systems consist of concrete, at grade tunnels, that allow access along the entire route. These systems can be used for all the listed media. The CST design should include the following:

- Detailed piping layouts showing all support locations
- Clearances inside the trench system, insulation, and jacketing selection

- Concrete trench wall and floor design (cast-in-place)
- Concrete top design (precast or cast-in-place) complete with lifting devices
- All road crossings
- Grading to keep groundwater from ponding over the trench system
- Sealant types and locations
- System slope 1 in./20 ft [25 mm/6096 mm] minimum, to ensure the trench floor will drain to the valve manholes)
- Vent locations

All drains and traps must be located in the valve manholes. Vents may be located in the trench system only if access is provided to them with manhole lids poured in the trench top.

The use of manholes in these systems to provide housing for drains and traps must be evaluated concerning confined space entry provisions.

7.10.2.3 Buried Conduit (preapproved type)

Due to many premature failures, buried conduit (preapproved type) is the last choice in buried distribution systems. These systems consist of insulated steel carrier pipe enclosed in a drainable and dryable steel conduit. These systems are not preferred except in an unusual situation that precludes the use of any other system (e.g., flood plain areas).

Manufacturer's Responsibility

Buried conduit systems are transported to the site in factory assembled sections. The manufacturer is responsible for the design of pipe supports, expansion compensation devices, end seals, insulation types, conduit design, and universal protection of the conduit. The manufacturer must submit expansion stress calculations for the designer to review compliance with project specifications.

Designer's Responsibility

The designer is responsible for all the general design considerations listed previously and should also include the design of the buried conduit system's penetration into the concrete valve manhole and the detailed routing of the system on the site.

The use of manholes in these systems to provide housing for drains and traps must be evaluated concerning confined space entry provisions.

7.10.2.4 Buried Conduit (not preapproved type)

BC systems (not preapproved type) consist of an insulated metallic or nonmetallic carrier pipe covered by a nonmetallic conduit. Due to the lower pressures and temperatures of these media, these systems have proven effective.

BC systems (not preapproved type) are similar to the preapproved buried conduit in that these systems are delivered to the site in factory assembled sections. However, the designer has less control with the not preapproved system. The designer chooses the items listed for general design considerations, and, in addition, provides detailed piping layouts, insulation type and thickness, conduit selection, carrier pipe selection, and valve manhole entrances.

The use of manholes in these systems to provide housing for drains and traps must be evaluated concerning confined space entry provisions.

7.10.3 Design of Heat Distribution Systems

The design of heat distribution systems includes, but is not limited to the following:

- Mechanical—expansion compensation, piping system design (fittings, valves, insulation), equipment selection, equipment sizing, and pipe sizing and routing
- Structural—reinforced concrete design, pipe supports, valve manhole design, and other miscellaneous structural designs
- Electrical—electrical service to equipment and controls, and universal protection (if required)
- Civil—excavation and backfill, grading, road crossings for buried systems, area drainage design, system plans and profiles, and site coordination to ensure system integrity (especially for CST) fits into the site properly

7.10.4 Existing System Capacity

The designer must determine if the system has adequate capacity to tie into the existing heat distribution system. The designer must also determine if the connecting points for the existing lines have adequate hydraulic capacity (are large enough) to satisfactorily handle the new loadings under variable operational scenarios.

Each installation should have hydraulic analysis data to indicate what the new loading impact is on the existing system. This information must be provided by the designer. The designer must update the hydraulic analysis, while considering possible future expansion impacts, as part of any new system design.

7.10.5 General Design Considerations

The following general design considerations should always be considered:

- Survey—A survey in the location of the distribution system must be done complete with soil borings and information on groundwater, soil types, and soil resistivity. The survey data should be noted.
- Utilities—A utility investigation must identify all existing utilities within a minimum utility corridor of 25 ft (7.6 m) of the new distribution system (including information on type, piping material, size, and depth). This investigation includes the engineering determination of where to connect the new distribution system to the existing system. All new connections must be at or near existing system anchor points to avoid damage to the existing utility system.
- Pipe sizing—All new pipes must be sized in accordance with prescribed engineering design procedures. Minimum line sizes for any system should be 1.5 in. (38 mm) (nominal). The use of better performing pipe materials for specific trench soils should be a consideration.

- Expansion—Expansion compensation calculations are necessary to ensure the new lines are properly designed under the engineering allowable values for stresses, forces, and moments. A computer finite element analysis program can be used to determine these values. Only loops and bends are to be used for expansion compensation. No expansion joints should be permitted in the design and installation.
- Valve manholes—Concrete valve manholes must be completely designed including structural grated or concrete covers, internals (including valves, traps and drip legs), clearances, and reinforced concrete design.
- Drainage—All valve manholes must either be gravity drained to an existing storm drain line with backflow protection or to a remote sump basin complete with duplex sump pumps, which discharge to an existing storm drain line or to grade.
- Grading—Regardless of the system, grading must be designed to prevent groundwater from entering the valve manholes.
- Plan/profile—Plans and profiles should be drawn for all systems showing, at a minimum:
  —System routing and piping slope elevations
  —System stationing
  —All existing utility and other major interferences (depths if known)
  —All adjacent roads and buildings clearly labeled
  —Current types of surface conditions along the new utility corridor (asphalt, grass)
  —Both new and existing grade contour lines (plan)
  —Exact support locations for the new utility system
  —Dimensioning (consistent English or metric units) to ensure accurate utility routing

7.10.6 Identification

Provide a brass name tag for each valve and temperature control device installed in all mechanical systems.

All exposed or concealed piping in accessible spaces should be identified with color-coded bands and titles in accordance with American National Standards Institute (ANSI) Standard A13.1, Scheme for Identification of Piping Systems.

- Pipes in buildings are categorized as pipes related to
  —Fire protection systems
  —Critical piping in essential and hazardous facilities
  —All other piping
- All water pipes for fire protection systems in seismic zones 1, 2, 3, and 4 will be designed under the provisions of the current issue of the Standard for the Installation of Sprinkler Systems of the National Fire Protection Association (NFPA No. 30). To avoid conflict with these NFPA recommendations, the criteria in the following subsection are not applicable to piping expressly designed for fire protection.
- Ductwork in buildings is categorized as
  —Critical ductwork in essential and hazardous facilities
  —All other ductwork

Consistent system identification provides a basis for future communication to maintenance and operations personnel, users of the system, and emergency providers.

7.11 THERMAL INSULATION OF MECHANICAL SYSTEMS

This section contains requirements for the insulation of mechanical systems, including insulation of plumbing systems and equipment, roof storm drain system, hot water piping systems and equipment, chilled water piping and equipment, and the insulation of the duct systems.

- Air-conditioning return ducts located in ceiling spaces used as return air plenums do not require insulation.
- Hot water and chilled water circulating pumps should not be insulated.
- Provide reusable insulation covers at
  —All check valves
  —Control valves
  —Strainers
  —Filters
  —Any other piping component requiring access for routine maintenance
- Insulation exposed to the weather or possible physical damage should be covered by appropriate metal jackets. All piping with metal jackets should be identified on the drawings.

The use of insulation must also be evaluated regarding the potential for leakage from piping and/or condensation, which renders insulation a potential site of biological amplification.

7.12 PLUMBING SYSTEM

The plumbing system consists of the water supply distribution system; fixtures and fixture traps; soil, waste, and vent piping; storm water drainage; and acid and industrial waste disposal systems. It extends from connections within the structure to a point 5 ft (1.5 m) outside the structure. The design of all plumbing must comply with the most current National Standard Plumbing Code, unless otherwise stated.

- Pipe materials for the domestic water system should be specified as nonferrous.
- Underground water pipes must be installed below the recognized frost line or insulated to prevent freezing.
  —Service lines enter the building in an accessible location, and when entering through the floor, a displacement type water entrance should be provided.
  —When the incoming pressure of water supply exceeds the water pressure necessary for proper building operation by 10 psig (68.9 kPa), a pressure-reducing valve must be provided.

7.12.1 Piping Run

Piping runs should be designed to minimize interference with ordinary movement of personnel and equipment.

- The water supply piping is distributed throughout the building, with water mains generally running near the ceiling of the lowest floor.

Neither water nor drainage piping should be located over electrical wiring or equipment unless adequate protection against water intrusion (including condensation) damage has been provided. Insulation alone is not adequate protection against condensation.

- Water and waste piping should not be located in exterior walls, attics, or other spaces wherever a danger of freezing exists. Where piping is to be concealed in wall spaces or pipe chases, such spaces should be checked to insure that clearances are adequate to properly accommodate the piping. Water piping should be designed for a maximum flow velocity of 8 ft/s.

Pipe chases and collocation of piping must be evaluated for accessibility and the potential for hosting contaminant repositories if leakage occurs. Both biological and chemical risk should be evaluated, particularly for spaces where small leaks may go unnoticed.

- Cross connections between water supply piping and waste, drain, vent, or sewer piping are prohibited.
  —Piping should be designed so that a negative pressure in the water supply pipe and/or a stopped-up waste, drain, vent, or sewer pipe will not cause backflow of wastewater into the water supply piping.
  —Single check valves are not considered adequate protection against wastewater backflow.

7.12.1.1 Back-Siphonage

The supply outlet connection to each fixture or appliance that is subject to back-siphonage of nonpotable liquids, solids, or gases must be protected in accordance with the National Standard Plumbing Code.

Depending on the severity of the backflow situation, an air gap, atmospheric vacuum breaker, double check valve assembly, or reduced-pressure device may be required. Severe backflow situations may include systems connected to boilers or converters containing glycol mixtures, which should require a reduced-pressure device.

- Air gaps will conform to the National Standard Plumbing Code.
- Double-check valve assemblies, reduced-pressure assemblies, atmospheric (non-pressure) type vacuum breakers, and pressure type vacuum breakers will be tested, approved, and listed by the Foundation for Cross-Connection Control and Hydraulic Research.
- Atmospheric type vacuum breakers, hose connection vacuum breakers, and backflow preventers with intermediate atmospheric vents will be in accordance with American Society of Sanitary Engineering (ASSE) Standards 1001, 1011, and 1012.
- Servicing stop valves should be installed in all water connections to all installed equipment items, as necessary for normal maintenance or replacement, and should be shown on the drawings, except when called for in project specifications.
- Water conservation fixtures (low-flow type) conforming to the guide specifications will be used for all toilets, urinals, lavatory faucets, and shower heads, except where the sewer system will not adequately dispose of the waste material on the reduced amount of water.
- Commercially available water hammer arresters should be provided at all quick closing valves, such as solenoid valves, and will be installed according to manufacturers' recommendations. Vertical capped pipe columns are not permitted.

- Electric, refrigerated water coolers should be used for all drinking water requirements, except in hazardous areas per NEC Article 500. Refrigerant R-12 should be not be used if possible; use Refrigerant R-22 or R-134a instead.
- Freeze-proof wall hydrants with vacuum breaker backflow preventers should be located on outside walls so that, with no more than 100 ft (30.5 m) of garden hose, an area can be watered without crossing the main building entrances.
- Emergency showers and eyewash stations should be provided where hazardous materials are stored or used or as required by the installation facility manager and should be installed in accordance with ANSI Standard Z385.1, the current version.
  —Where the eyes or body of any person may be exposed to injurious corrosive materials, an emergency shower and eyewash station should be provided conforming to the ANSI Standard Z385.1.
  —In accordance with ANSI Standard Z385.1, a heated water system should provide tempered water (60–100°F [16–38°C]) for a 15-min duration at the flow rate required by the installed shower/eyewash.
- The domestic hot-water heating energy source should be steam, HTHW, natural gas, fuel-oil, or electricity. The use of electricity should be avoided if possible. Electricity is permitted for point-of-use water heaters only. Domestic hot-water design temperatures should be 120°F (49°C).
- Criteria determining the need for circulating pumps as shown in the American Society of Heating, Refrigerating, and Air-Conditioning (ASHRAE) Handbook *HVAC Applications* must be followed. Within buildings operated on a nominal 40-h week or on a nominal two-shift basis (either a 5-day or a 7-day week), a design should include installation of a clock or other automatic control on the domestic hot-water circulating pumps to permit operation only during periods of occupancy, plus 30 min before and after working hours.
- Floor drains should be provided in toilet rooms with three or more toilets. Provide floor drains in shower drying areas serving two or more showers. Provide enough floor drains in utility and boiler rooms to avoid running equipment drain pipes above the floor.
- The selection of pipe and fitting materials for acid waste and vent applications will be based on the type, concentration, and temperature of acid waste to be handled. Acid neutralization tanks should be provided for all acid waste drainage systems.

All acid waste systems must be evaluated for potential worker exposure in case overhead leaks occur. Collocation of caustic and thermal hazard lines must also be evaluated for increased hazard.

- Where feasible, provide circuit vents in a concealed space to a main vent through the roof in lieu of an excessive number of individual roof vents. Waste and vent piping should be concealed unless otherwise specifically instructed.
- Storm drainage will include roof drains, leaders, and conductors within the building and to a point 5 ft (1.5 m) outside the building. Roof drainage systems will be designed in accordance with rainfall intensity-frequency data in the National Standard Plumbing Code.

7.13 COMPRESSED AIR SYSTEM

Low-pressure compressed air systems have a maximum design operating pressure of 200 psig (1378 kPa), including piping and compressors. Compressed air systems must be designed in accordance with ASME B19.1-1985 and B19.1a-1985, Safety Standards for Compressor Systems, the current version.

7.13.1 Compressor Selection and Analysis

A central compressed air system will be utilized to serve multiple points of use. Compressors and all accessories will conform to American Society of Mechanical Engineers (ASME) B19.1 and B19.3; ASME Boiler and Pressure Vessel Code Section VIII, PTC-9 & PTC-10; and the Instrumentation, Systems, and Automation Society (ISA) S7.3, as applicable.

- Where lubricating oils cannot be tolerated at the point of use, oil-free air compressors will be used.
- For isolated areas where oil-free air is required in a nonoil-free compressed air system, coalescing filters may be used to remove solids, moisture, and oil from the airstream in lieu of an oil-free compressor.

Comparisons of such items including, but not limited to, brake horsepower (bhp) per 100 CFM (47.2 l/s), unloaded horsepower, expected compressor life, and expected operation and maintenance costs should be made between the different types of compressors before a final selection is made.

7.13.2 Compressor Capacity

The total air requirement will not be based on the total of individual maximum requirements, but will based on the sum of the average air consumption of air-operated devices. Determination of the average air consumption is based on the concept of load factor (the ratio of actual air consumption to the maximum continuous fully loaded air consumption).

The Compressed Air and Gas Institute (CAGI) *Compressed Air and Gas Handbook* explains the procedure for using load factors to determine compressor capacity. After making the calculation, add 10% to the estimated consumption for air leakage. The total air is the calculated compressor capacity required for design. More capacity may be added to allow for future growth of the facility or serviced area over the next 2 years.

7.13.3 Compressor Location and Foundations

Locate compressors within a ground floor utility or mechanical equipment room with adequate space to permit easy access for cleaning, inspection, and any necessary dismantling. Adequate aisle space is also needed between equipment for normal maintenance as well as for equipment removal and replacement. Foundations that are isolated from the building structure should be provided for all compressors.

7.13.4 Makeup Air

For large air compressors located in closed mechanical rooms, a wall opening should be provided for makeup air. Exterior wall openings should be provided with louvers and motorized dampers.

7.13.5 Compressed Air Outlets

A ball valve, a pressure-reducing valve, a filter, and a quick-disconnect coupling should be provided at each compressed air outlet.

7.13.6 Refrigerated Dryer

Some compressed air applications require moisture removal in addition to that provided by an aftercooler. Such commercial applications include paint spraying, sandblasting, the use of air-operated tools and devices, pneumatic automatic temperature controls, lines run outside in cold or subfreezing locations, and lines passing through cold storage rooms. Where moisture removal is required, provide a refrigerated type air dryer located downstream from the compressor initial exhaust duct area and prior to discharge to the environment.

7.14 AIR SUPPLY AND DISTRIBUTION SYSTEM

The design of all systems must comply with the ASHRAE handbook and to the requirements of NFPA Standards Nos. 90A, 90B, and 91.

7.14.1 Basic Design Principles

All designs will be based on the following basic principles:

- Interior design conditions, including temperature, humidity, filtration, ventilation, and air changes, will be suitable for the intended occupancy.
- The designer will evaluate all energy conservation items that appear to have potential for savings, such as heat recovery for HVAC and service water heating, economizer cycles, and plastic door strips for load docks, and will include those items in the design that are life cycle cost-effective.
- The design will be as simple as possible.
- Adequate space will be provided for maintenance access to ancillary equipment such as filters, coils and drain pans, and strainers.
- To the extent practical, system airflow will be minimized. Integrated air-conditioning and lighting systems will be used whenever the general lighting level is 100 fc or greater.
- Recovered heat will be used for reheat where possible.

7.14.2 Temperature Settings

HVAC sequence of control should include procedures for personnel to reset HVAC control settings in occupied zones from 76°F (24°C) up to 78°F (26°C) if future energy conservation actions are required.

- The design relative humidity will be at least 50% or the design temperature equal to the outside air dew point design temperature, whichever is less.
- The indoor design specific humidity will not exceed the outdoor design specific humidity; otherwise the indoor design relative humidity will be 50%.
- The indoor design temperature provided by evaporative cooling or comfort *mechanical* ventilation will be 80°F.

7.14.3 Air-Conditioning Loads

- Air-conditioning loads should be calculated using ASHRAE methods. The designer should plot the following on a psychometric chart:
  —Entering and leaving air temperature conditions for the coil
  —Expected room conditions
  —Outside air conditions for each air system

7.14.4 Infiltration

Where acceptable, air distribution systems for the central HVAC systems will be designed to maintain a slight positive pressure within the area served to reduce or eliminate infiltration.

7.14.5 Outdoor Air Intakes

Outdoor air intakes will be located in areas where the potential for air contamination is lowest. This is a common design problem that routinely needs correction. Basic guidelines for air intake location include the following:

- Maximize distance between the air intakes and all cooling towers, plumbing vents, loading docks, and traffic stations
- Maintain a minimum distance of 30 ft (9.2 m) between air intakes and exhausts— more if possible
- Locate air intakes and exhausts on different building faces

7.14.6 Filtration

For administrative facilities, commercial facilities, and similar occupancies where IAQ is of primary concern, the combined supply air, including return and outside air, should be filtered. Filtration uses a combination of 25 to 30% efficient prefilters and 80 to 85% efficient final filters as determined by the dust spot test specified in ASHRAE Standard 52.

Due to the decrease in system airflow as the pressure drop across the filters increases, fans should be sized for the "dirty" filter condition. This sizing will ensure that the fan has adequate capacity to deliver the design airflow as the filter becomes loaded. To ensure that this fan capacity is "available," test and balance criteria need to estimated appropriately.

7.14.7 Economizer Cycle

- Provide outside air "temperature economizer cycle" for comfort air-conditioning or equipment cooling only when humidity control is not required.
- Provide economizer cycle only on systems greater than 3000 CFM (1,416 l) that are operated 8 or more hours per day.
- Enthalpy control for the economizer cycle should not be provided.

7.15 DUCTWORK DESIGN

All ductwork for heating/ventilating-only systems should be insulated where future air-conditioning of the building is anticipated.

- Return air ductwork should be routed into each area isolated by walls, which extend to the roof structure. The designer should not use transfer ducts or openings.
- The use of round or oval prefabricated ductwork is recommended. Round/oval prefabricated ductwork reduces leakage and friction losses and reduces the amount of conditioning and fan energy required. The additional material cost for round/oval prefabricated ductwork would be at least partially offset by cost and time savings.

7.15.1 Variable Air Volume (VAV) Systems

VAV air handling systems and their associated HVAC control systems, due to complexity, require more critical and thorough design. When VAV is selected over other types, the following questions must be discussed during design:

- Were other HVAC systems considered and why were they not selected?
- Was a constant volume system with VAV bypass boxes considered?
- How will outside ventilation air be controlled during periods of low cooling loads?
- How will adequate heating be provided along outside walls and perimeter zones, including the need for supplemental baseboard heat?
- Was a multizone system with space discriminator reset of hot and cold deck temperatures or a single zone system with space discriminator control of supply air temperature considered in the design process?

7.15.2 Special Criteria for Humid Areas

The criteria described in this section must be used in the design of air-conditioned facilities located in areas where the

- Wet bulb temperature is 67°F (19°C) or higher for over 3000 h and outside design relative humidity of 50% or higher
- Wet bulb temperature is 73°F (23°C) or higher for over 1500 h and the outside design relative humidity is 50% or higher, based on 2.5% dry bulb and 5.0% wet bulb temperatures

Air-conditioning will be provided by an all-air system. The system may consist of a central air-handling unit with chilled water coils or a unitary direct expansion-type unit capable of controlling the dew point of the supply air for all load conditions. The following systems should be considered:

- Variable volume constant temperature
- Bypass variable air volume
- Variable temperature constant volume
- Terminal air blenders

In addition to life-cycle cost considerations, system selection will be based on the capability of the air-conditioning system to control the humidity in the conditioned space continuously under full load and part load conditions.

- *System selection should be supported by an energy analysis computer program that will consider the latent heat gain due to*

—Vapor flow through the building structure

—Air bypassed through cooling coils

—Dehumidification performance of the air-conditioning system under varying external and internal load conditions

- Low sensible loads and high latent loads (relatively cool cloudy days) will, in some cases, raise the inside relative humidity higher than desired. If analysis indicates that this condition will occur, reheat must be used in the design selection.
- Room fan coil units will not be used unless dehumidified ventilation air is supplied to each unit or separately to the space served by the unit and positive pressure is maintained in the space.
- Draw-through type air-handling units will be specified to use the fan energy for reheat. The air distribution system will be designed to prevent infiltration at the highest anticipated sustained prevailing wind.
- Outside air will be conditioned at all times through a continuously operating air-conditioning system.
- The supply air temperature and quantity and chilled water temperature will be based on the
  —Sensible heat factor
  —Coil bypass factor
  —Apparatus dew point
- The 1% wet bulb temperature will be used in cooling calculations and equipment selections.
- Closets and storage areas should be either directly air-conditioned or provided with exhaust to transfer conditioned air from adjacent spaces.
- Where reheat is required to maintain indoor relative humidity below 60%, heat recovery, such as reclamation of condenser heat, should be considered in life cycle cost analysis.
- Economizer cycles will generally not be used due to the high moisture content of outside air.

7.15.3 Evaporative Cooling

Evaporative cooling may be used where the facility in question is eligible for air-conditioning, and evaporative cooling can provide the required indoor design conditions based on the appropriate outdoor design conditions. For special applications where close temperature or humidity control is required, two-stage evaporative cooling or indirect evaporative cooling should be considered in life-cycle cost analysis as a supplement to, not in lieu of, a primary cooling system.

7.16 VENTILATION AND EXHAUST SYSTEMS

The design of all systems should comply with the ASHRAE handbook, ASHRAE Standard 62, and the requirements of NFPA Standards Nos. 90A, 90B, and 91. Motorized low-leakage dampers, with blade and jamb seals, should be provided at all outside air intakes and exhausts.

7.16.1 Supply and Exhaust Fans

Exterior wall and roof-mounted supply or exhaust fans should be avoided; connect interior fans with ductwork and louvers.

Except for interior wall-mounted propeller units, all fans should be centrifugal type and connected directly to weatherproof louvers or roof vents via ductwork.

- Fans larger than 2000 CFM (944 l/s) should be provided with V-belt drives.
- Care should be taken to prevent the noise level generated by exhaust fans and associated relief louvers from being transmitted to the exterior of the building. Any in-line fans located outside the main mechanical and electrical areas should be provided with acoustical enclosures to inhibit noise transmission to the adjoining occupied spaces, depending on occupant use.

Where possible, exhaust fans in all buildings in housing, recreational, hospital, and administrative areas should be of the centrifugal type, discharging through louvers in the side wall of the building using ductwork, as necessary. Roof-mounted fans of the low-silhouette type may be used.

Centrifugal type roof exhausters should be used in shop, flight line, or warehouse areas. Where exhaust ventilating fans or intakes are provided in buildings, a positive means (gravity dampers are not acceptable) of closing the fan housing or ducts should be provided to prevent heat loss in cold weather, except as prohibited by NFPA Standard 96.

7.16.2 General Items

Incorporate the following:

- Ventilation for VAV systems will ensure proper ventilation rates at low and high system airflow.
- Year-round supply (makeup) air should be provided to equal the total quantity of all exhaust hoods.
- Where desirable, incorporate a purge mode into system design. This mode could be used, for example, to purge the building with outside air during off-hours or to purge the affected zone during building maintenance, such as painting.
- The toilet rooms and janitor closet should be exhausted at a rate of 2 CFM/ft^2 (10 l/s/m^2) by insulated in-line fans to maintain a negative room pressure. The required makeup air for the exhaust system should be from undercut doors or, if necessary, through door grilles. Exhaust registers, in lieu of grilles, should be provided in areas with rigid ceilings.
- Shower areas have a 2.5 CFM/ft^2 (13 l/s/m^2) exhaust rate to maintain a negative room pressure.
- Where practical, photocopiers, laser printers, and print equipment should be located in a separate room. Copy rooms with photocopiers and laser printers should not be directly conditioned, but should be maintained at a negative pressure relative to adjacent areas by exhausting air from these adjacent areas directly to the outdoors. All conditioned supply air to the room should be exhausted and not returned to the air-handling unit system due to contaminants.
- Mechanical and electrical equipment rooms should be ventilated and cooled with outside air by thermostatically controlled fans set to operate when the temperature exceeds 85°F (29°C).
- The boiler room should be ventilated and cooled with outside air at a minimum rate of 20 air changes/h by a thermostatically controlled supply or exhaust fan set to operate when temperature exceeds 85°F (29°C). Supply fans should be used when atmospheric burners are permitted.

- The fire protection room should be ventilated and cooled with outside air by a thermostatically controlled fan set to operate when the temperature exceeds 85°F (29°C).
- Provide exhaust fans in laundry rooms sized for a minimum of 3-min air changes.
- Automotive maintenance shops must be provided with a suitable engine exhaust ventilating system. General ventilation should be provided at 1.5 CFM/ft^2 (8 l/s/m^2) of outside air.
- Battery rooms should be ventilated at a rate of four air changes per hour.

7.17 TESTING, ADJUSTING, AND BALANCING OF HVAC SYSTEMS

All test and inspection reports and the following should be completed before starting the distributed control system (DCS) field test.

- Testing, adjusting, and balancing should be performed by an independent firm using certified technicians under the direct supervision of a registered engineer.
  —Technicians should be certified by the National Environmental Balancing Bureau (NEBB) or by the Associated Air Balance Council (AABC).
  —The firm should select AABC MN-1, or NEBB-01 as the standard for testing, adjusting, and balancing the mechanical systems.
- Air-handling unit filters should be artificially loaded during testing and balancing operations. Air-handling unit airflow should be set for maximum with the filters fully loaded.

7.18 VENTILATION ADEQUACY

Ventilation systems are designed to protect the health of individuals by removing physical and chemical stresses from the workplace. To ensure that these ventilation systems are operating effectively, ventilation flow rates may require periodic checking. Manufacturer's specifications and applicable guidelines for specific types of equipment and applications are utilized to ensure the proper operation of local exhaust ventilation.

Capture velocities may vary depending on contaminant size, generation rate, air currents, and other variables. Each local exhaust ventilation system must be independently evaluated to determine adequate operating parameters.

- Local exhaust ventilation systems without a static pressure manometer should have performance evaluations conducted annually.
- Local exhaust ventilation systems with a static pressure manometer must have performance evaluations every 3 years.

7.19 LABORATORY FUME HOOD PERFORMANCE CRITERIA

Face velocity measurements must be 90–150 fpm with the fume hood sash fully opened, unless superseded by manufacturer's specifications or other applicable guidelines. Readings should be obtained and recorded every 3 months for fume hoods both with and without a static pressure manometer. If the static pressure deviates ±10%, the fume hood will be inspected and reevaluated. Inspections need to be documented.

7.20 FLOW HOODS

Flow hoods measure air velocities at air supply or exhaust outlets.

7.20.1 Calibration

No field calibration is available for flow hoods. Periodic calibration by a laboratory is essential.

7.20.2 Maintenance

Flow hoods typically require little field maintenance other than battery-pack servicing and zero balancing of analog scales. (Check the applicable manufacturer's manual.)

7.21 THERMOANEMOMETERS

Thermoanemometers monitor the effectiveness of ventilation by measuring air velocities.

7.21.1 Calibration

No field calibration is available for thermoanemometers. Periodic calibration by a laboratory is essential.

7.21.2 Maintenance

Thermoanemometers typically require little field maintenance other than battery-pack servicing and zero balancing of analog scales. (Check the applicable manufacturer's manual.)

7.22 OTHER VELOMETERS

Other velometers include rotating-vane and swinging vane velometers.
Note: Barometric pressure and air temperature should be noted when using air velocity meters.

CHAPTER 8

Area Monitoring and Contingency Planning

All sites and facilities with over ten employees are required to have contingency plans; this chapter discusses the requirements for air monitoring in such facilities. Air monitoring for sites known to be hazardous is also discussed, again with real-world emphasis and examples.

8.1 AREA OF INFLUENCE PERIMETER

8.1.1 Evaluation of Hazardous Waste/Chemical Risk Sites

Site characterization provides the information necessary to identify site hazards and select worker protection methods. The more accurate, detailed, and comprehensive the information available about a site, the more the protective measures will be tailored to the actual hazards workers may encounter. Site characterization generally proceeds in two phases:

1. Obtain as much information as possible before site entry so hazards can be evaluated and preliminary controls established to protect initial entry personnel.
2. Initial information-gathering missions will focus on identifying all potential or suspected conditions that may present inhalation hazards, which are immediately dangerous to life or health (IDLH), and any other conditions that may cause death or serious personal harm.

8.1.2 Off-Site Characterization before Site Entry

Before going to the hazardous waste/chemical risk site, the off-site characterization will be used to develop a site safety and health plan. The site safety and health plan addresses the work to be accomplished and prescribes the procedures to protect the safety and health of the entry team.

In the site safety and health plan, after careful evaluation of probable site conditions, priorities will be established for hazard assessment and site activities. Because team members may enter a largely unknown environment, caution and conservative actions are appropriate, which should be reflected in the site safety and health plan for the hazardous waste/chemical risk site.

8.1.2.1 Interview/Records Research

Collect as much data as possible before any personnel go onto the hazardous waste/ chemical risk site. When possible, obtain the following information:

- On-site conditions; exact location of the site
- Detailed description of the activities that have occurred or are occurring at the site
- Duration of the activity
- Meteorological data, e.g., current weather and forecast, prevailing wind direction, precipitation levels, temperature profiles
- Terrain, e.g., historical and current site maps, site photographs, U.S. Geological Survey topographic quadrangle maps, land use maps, and land cover maps
- Geologic and hydrologic data
- Habitation, including population center, population at risk, and ecological receptors
- Accessibility by air and roads
- Pathways of contaminant dispersion
- Present status of response (Who has responded?)
- Hazardous substances involved and their chemical and physical properties
- Historical records search:
    - —Company records, receipts, logbooks, or ledgers
    - —Records from state and federal pollution control regulatory and enforcement agencies, state attorney general offices, state occupational safety and health agencies, or state fire marshal offices
    - —Waste storage inventories, manifests, or shipping papers
    - —Generator and transporter records
    - —Water department and sewage district records
    - —Local fire and police department records
    - —Court records
    - —Utility company records
- Interviews with personnel and their families (Verify all interview information, if possible. **Note:** Issues of confidentiality may be involved.)
- Interviews with nearby residents (Note possible site-related medical problems and verify all information from interviews, if possible.)
- Media reports (Verify all information from the media, if possible.)

8.1.3 On-Site Survey

During on-site surveys site entry will be restricted to reconnaissance personnel. Particular attention will be given to potentially IDLH conditions. The purpose of the on-site survey is to verify and supplement information gained from the off-site characterization.

The composition of the entry team depends on the site characteristics, but should always consist of at least four persons. Two workers will enter the site [exclusion zone (EZ) and contamination reduction zone (CRZ)]. The other two persons will remain in the support zone (SZ), suited up in the same PPE as the personnel in the EZ/CRZ. The support personnel are on alert in case of emergency and will be prepared to enter immediately if an emergency occurs.

Ongoing monitoring will provide a continuous source of information about site conditions. Site characterization is a continuous process. During each phase information will be collected and evaluated to define the hazards present at the site. In addition to the formal information gathering described here, all site personnel will be constantly alert for new information about site conditions.

- Areas on-site or at facilities that may be subject to chemical exposures need to be monitored both to determine potential worker exposures and off-site effects.
- Monitoring must be conducted before site entry at uncontrolled hazardous waste sites to identify IDLH conditions, such as oxygen-deficient atmospheres and areas where toxic substance exposures are above permissible limits.
- Accurate information on the identification and quantification of airborne contaminants is useful for
  —Selecting PPE
  —Delineating areas where protection and controls are needed
  —Assessing the potential health effects of exposure
  —Determining the need for specific medical monitoring

After a hazardous waste cleanup operation begins, periodic monitoring of those employees who are likely to have higher exposures must be conducted to determine if they have been exposed to hazardous substances in excess of the OSHA PELs. Monitoring must also be conducted for any potential IDLH condition or for higher exposures that may occur as a result of new work operations.

8.1.3.1 Potential IDLH Conditions

Visible indicators of potential IDLH and other dangerous conditions include the following:

- Containers or tanks that will be entered
- Enclosed spaces such as buildings or trenches that will be entered
- Potentially explosive or flammable situations indicated by bulging drums, effervescence (bubbles like carbonated water), gas generation, or instrument readings
- Extremely hazardous materials e.g., cyanide, phosgene, or some radiation sources
- Vapor clouds
- Areas where biological indicators (such as dead animals or vegetation) are located

8.1.3.2 Perimeter Reconnaissance

Research previous soil surveys, ground-penetrating radar and manometer data, and air sampling and monitoring data. Monitor atmospheric conditions and airborne pollutants. Such data are not a definitive indicator of the site conditions, but assists in the preliminary evaluation.

Perimeter reconnaissance of a site will involve the following actions:

- Develop a preliminary site map that shows the locations of buildings, containers, impoundments, pits, ponds, existing wells, and tanks.
- Review historical and recent aerial photographs. Note any of the following:
  —Disappearance of natural depressions, quarries, or pits
  —Variation in revegetation of disturbed areas

    —Mounding or uplift in disturbed areas or paved surfaces or modifications in
      grade
    —Changes in vegetation around buildings or anywhere else on-site
    —Changes in traffic patterns at the site
    —Labels, markings, or placards on containers or vehicles
    —Amount of deterioration or damage to containers or vehicles
    —Biologic indicators, e.g., dead animals or plants, discolored soils and/or plants,
      or the total lack of vegetation in some areas
    —Unusual conditions, e.g., clouds, discolored liquids, oil slicks, vapors, or other
      suspicious substances
    —Toxic substances
    —Combustible and flammable gases or vapors
    —Oxygen deficiency
    —Ionizing radiation
    —Unusual odors
- Collect and analyze off-site samples, including the following:
    —Soil
    —Drinking water
    —Groundwater
    —Site runoff
    —Surface water

8.1.3.3 On-Site Survey

After entering the site, the entry personnel will gather the following information as
quickly and carefully as possible:

- Monitor the air for IDLH and other conditions that may cause death or serious harm
  (combustible or explosive atmospheres, oxygen deficiency, toxic substances, etc.).
- Monitor for ionizing radiation (survey for alpha, beta, and gamma radiation).
- Look for signs of actual or potential IDLH or other dangerous conditions. Any
  indication of IDLH hazards or other dangerous conditions will be regarded as a
  sign to proceed with caution, if at all. If the site safety and health plan does not
  cover the conditions encountered, exit the site and reevaluate the plan. Exercise
  extreme caution in conducting site surveys when such hazards are indicated. If
  IDLH or other dangerous conditions are not present, or if proper precautions can
  be taken, continue the survey after field modifying the site safety and health plan.
- Survey the on-site storage systems and contained materials. Note the types of
  containers, impoundments, or other storage systems present, such as
    —Paper or wood packages
    —Metal or plastic barrels or drums
    —Underground tanks
    —Aboveground tanks
    —Compressed gas cylinders
    —Pits, ponds, or lagoons
- Note the condition of the waste containers and storage systems, such as
    —Structurally sound (undamaged)
    —Visibly rusted or corroded
    —Leaking
    —Bulging

- Note the types and quantities of material in containers, such as labels on containers indicating corrosive, explosive, flammable, radioactive, or toxic materials
- Note the physical condition of the materials:
  —Gas, liquid, or solid
  —Color and turgidity
  —Chemical activity, e.g., corroding, foaming, or vaporizing
  —Conditions conducive to splash or contact
- Identify natural wind barriers:
  —Buildings
  —Hills
  —Aboveground tanks
- Determine potential dispersion pathways:
  —Air
  —Biologic routes, e.g., animals and food chains
  —Groundwater
  —Land surface
  —Surface water

If necessary, use one or more of the following remote sensing or subsurface investigative methods to find buried wastes or contaminant plumes:

- Electromagnetic resistivity
- Seismic refraction
- Magnetometry
- Metal detection
- Ground-penetrating radar

Note any indicators that hazardous substances may be present, such as

- Dead fish, animals, or vegetation
- Dust or spray in the air
- Fissures or cracks in solid surfaces that expose deep waste layers
- Pools of liquid
- Foams or oils on liquid surfaces
- Gas generation or effervescence
- Deteriorating containers
- Cleared land areas or possible land-filled areas

Note any safety hazards. Consider the following:

- Condition(s) of site structures
- Obstacles to entry or exit
- Terrain homogeneity, e.g., smooth or uneven surfaces, depressions
- Terrain stability, e.g., signs of cave-in or unstable soils
- Stability of stacked material
- Reactive, incompatible, flammable, or highly corrosive wastes

Note land features. Note the presence of any potential naturally occurring skin irritants or dermatitis agents, such as poison oak, poison ivy, or poison sumac.

8.1.4 Chemical Hazard Monitoring

Once the presence and concentrations of specific chemicals or classes of chemicals have been established, the hazards associated with these chemicals will be determined by referring to standard reference sources for data and guidelines on toxicity, flammability, and other hazards.

Proper documentation and document control are important for ensuring accurate communication, ensuring the quality of data collected, preserving and providing the rationale for safety decisions, and substantiating possible legal actions.

Documentation can be accomplished by recording information pertinent to field activities, sampling analysis, and site conditions in any of several ways, including, but not limited to

- Logbooks
- Field data records
- Graphs
- Photographs
- Sample labels
- Chain-of-custody records
- Analytical records

Ensure all documents are accounted for when the project is completed. Each group that performs work at hazardous waste/chemical risk sites is responsible for setting up a document control system.

Document control will be assigned to one individual on the project team and will include the following responsibilities:

- Know the current location of documents (including sample labels).
- Record the location of each document in a separate document register so that any document can be easily located. (In particular, the names and assignments of site personnel with custody of documents will be recorded.)
- Collect all documents at the end of each work period.

8.1.4.1 Skin and Dermal Hazards

Information on skin absorption is provided in the ACGIH publication, *Threshold Limit Values for Chemical Substances and Physical Agents*, in OSHA standard 29 CFR 1910.1000, and in other standard references. These documents identify substances that can be readily absorbed through skin, mucous membranes, and/or eyes from either airborne exposure or from direct contact with a liquid. This information is qualitative and indicates whether a substance may pose a dermal hazard, but not to what extent. Thus, decisions made concerning skin hazards are necessarily judgmental, and more conservative protective measures will be selected.

Many chemicals, although not absorbed, may cause skin irritation at the point of contact. Signs of skin irritation range from redness, swelling, or itching to burns that destroy skin tissue. Standard references will be used to determine the level of personal protection necessary for hazardous waste/chemical risk site workers.

8.1.4.2 Potential Eye Irritation

Quantitative data on eye irritation are not always available. Where a review of the literature indicates that a substance causes eye irritation, but no threshold is specified, have a competent health professional evaluate the data to determine the level of protection necessary for hazardous waste/chemical risk site workers.

8.1.4.3 Explosion and Flammability Ranges

When evaluating the fire or explosion potential at a hazardous waste site, all equipment used should be explosion proof or intrinsically safe.

Where flammable or explosive atmospheres are detected, ventilation may dilute the mixture to below the LEL. Ventilation is generally not recommended if concentrations exceed the UEL because the mixture will pass through the flammable/explosive range as dilution occurs. Note: O_2/CGI readings may not be accurate when oxygen concentrations in air are less than 19.5%.

8.1.5 Monitoring

Because site activities and weather conditions change, an ongoing air monitoring program should be implemented after the hazardous waste/chemical risk site characterization has shown that the site is safe for the commencement of further hazardous waste/chemical risk work.

Ongoing atmospheric chemical hazard monitoring will be conducted using a combination of stationary sampling equipment, personnel monitoring devices, and direct-reading instruments used for periodic area monitoring.

Where necessary, routes of exposure (other than inhalation) will be monitored. Depending on the physical properties and toxicity of the hazardous waste/chemical risk site materials, areas outside the actual waste site may have to be assessed for potential exposures resulting from hazardous waste/chemical risk site work.

Monitoring also includes continual evaluation of any changes in site conditions or work activities that could affect worker safety. When a significant change occurs, the hazards should be reassessed. Some indicators of the need for such reassessments are as follows:

- Commencement of a new work phase
- Change in job tasks during a work phase
- Change in season
- Change in weather, e.g., high- versus low-pressure systems
- Change in ambient contaminant levels

Collect samples from the following:

- Air
- Drainage ditches
- Soil, e.g., surface and subsurface
- Standing pools of liquids

- Storage containers
- Streams, ponds, and springs
- Groundwater, e.g., upgradient, beneath site, downgradient

Sample for or otherwise identify

- Biological or pathological hazards
- Radiological hazards

8.1.6 Field Logbook Entries

Field personnel will record all hazardous waste/chemical risk site activities and observations in a field logbook (a bound book with consecutively numbered pages). To ensure thoroughness and accuracy, entries will be made during or just after completing a task. All document entries should be made in waterproof black ink, reproducible to four copies. Field logbook entries to describe sampling will include the following:

- Date and time entry
- Purpose of sampling
- Name, address, and organizational element of personnel performing sampling
- Name and address of the sampled material's producer
- Type of material, e.g., sludge, wastewater
- Description of the sampled material's container
- Description of sample
- Chemical components and concentrations
- Number and size of samples taken
- Sampling point description and location
- Date and time sample collected
- Difficulties experienced in obtaining sample, e.g., sample representative of the bulk of the material
- Visual references, e.g., maps or photographs of the sampling site
- Field observations, e.g., weather conditions during the sampling period
- Field measurements of material properties, e.g., explosiveness, pH, flammability

Note whether chain-of-custody records have been filled out for the samples.

Photographs can be an accurate, objective addition to a field worker's written observations. Record the following information for each photograph in the field logbook:

- Date, time, and name of site
- Name of photographer
- Location of the subject within the site by drawing a simple sketch or general orientation (compass direction) of the photograph
- General description of the subject
- Film roll and exposure numbers
- Camera, lens, and film type used

Provide sampling team members with serially numbered sample labels or tags:

- Tags assigned to each person will be recorded in the field logbook.
- Lost, voided, or damaged labels will be noted in the field logbook.

- Labels will be firmly affixed to the sample containers using either gummed labels or labels attached by a string or wire.

Label information will include the following:

- The unique sample log number
- Date and time collected
- Source of the sample, e.g., name, location, and type of sample
- Preservative(s) used, e.g., additions to the sample, special storage necessary
- Analysis required
- Name of collector
- Pertinent field data, e.g., weather conditions and temperature

In addition to supporting litigation, written records of sample collection, transfer, storage, analysis, and destruction help ensure analytical results are interpreted properly.

Chain-of-custody information must be included on a chain-of-custody record that accompanies the sample from collection to destruction.

8.1.7 Radiation Monitoring

To ensure that internal and external exposures to radiation are as low as reasonably achievable (ALARA), all radioactive materials must remain confined to designated work and storage locations; exposures resulting from the storage and use of these materials must be adequately known and controlled.

Because some forms of radiation cannot be detected by the human senses, these objectives can be met only through the routine use of instruments and devices specifically designed for the detection and quantification of radiation. Radiation-monitoring activities utilizing such devices generally assess either the extent and location of radiation hazards in an area or the exposure received by personnel.

8.1.7.1 Area Monitoring

Routine monitoring of radiation levels in areas where radioactive materials are stored or used is essential for ensuring the control of these materials and for managing personnel exposure. Such monitoring activities can generally be classified as either contamination surveys or exposure rate surveys. Contamination can be defined as radioactive material in an unwanted place.

8.1.7.2 Contamination Surveys

Depending upon the types and quantities of radioactive materials in use, contamination surveys may be made directly with portable survey instruments or indirectly (removable contamination survey, wipe, or swipe survey) by wiping surfaces (approximately 100 cm^2) with a filter paper and counting the wipes in a liquid scintillation system.

A direct contamination survey is performed using a meter and detector appropriate to the nuclides in use in the area. For example, in surveying for ^{32}P contamination, one would use a GM detector (probe); for ^{125}I use a thin-window NaI scintillation detector (probe). An ion chamber would not be appropriate for a contamination survey.

When surveying an area for contamination, check the meter before every use for proper operation using a suitable check source, then move the probe with a slow, steady motion over the area. The meter has an integrator circuit and will take time to properly respond. Meters should be equipped with audio circuits so a surveyor can discriminate a change in "click" rates and resurvey suspected "hot spots."

Removable contamination consisting of ^{3}H, ^{14}C, or ^{35}S is best detected through the use of wipes and liquid scintillation counting; beta emissions from these radionuclides have insufficient energy to be efficiently detected by portable survey instruments. Wipes may also be appropriate when attempting to detect contamination in areas with higher than background radiation levels. For example, the use of a GM survey meter to detect ^{32}P contamination on the lip of a hood would not be practical if radiation levels at that point were already elevated from ^{32}P stored within the hood.

When performing a contamination survey, move the probe slowly and steadily, as close as possible to the object to be monitored to allow the meter time to respond and to prevent air absorption from reducing the count rate.

When radiation levels in an area are normal background, portable survey instruments can be quite effective in detecting certain types of radioactive contamination. Most GM meters can detect ^{32}P with efficiencies exceeding 20%, and ^{125}I can be detected at efficiencies nearing 20% with a thin crystal (NaI) scintillation probe. All survey instruments are only as good as their maintenance. A portable survey meter must be calibrated every 6 months and verified before each use by monitoring a suitable check source.

8.1.7.3 Exposure Rate Surveys

In addition to contamination monitoring, it is also important to assess exposure rates resulting from the storage and use of relatively large quantities of high-energy beta or gamma emitters. This information is important in planning and evaluating the control of the factors of time, distance, and shielding for the particular situation in order to minimize personnel exposure. In most situations a properly calibrated GM meter can give a reasonable estimate of the exposure rate. An ion chamber will give the most accurate estimate of exposure and should be used whenever measuring exposures to determine regulatory posting, measuring exposure to determine the transport index of a package, or measuring exposures that are more than a few millirems.

8.1.7.4 Personnel Monitoring

State and federal regulations mandate that employers whose workers receive occupational exposure to radiation must advise the worker annually of the worker's exposure to radiation. All workers who might receive a radiation dose greater than 10% of the applicable value in Table 8.1 must be issued a suitable radiation-monitoring device. The readings from these devices are recorded by the employer for review by the state. These readings make up the individual's official exposure record.

There are a number of types of materials or devices that are used to assess an individual's cumulative external radiation exposure, collectively termed *dosimeters*. The most commonly used dosimeter is the film badge, which consists of a small piece of radiation-sensitive film placed in a special holder containing various filters. The film badge is worn by the radiation worker somewhere on the torso whenever working with or near radioactive materials emitting penetrating radiations (i.e., energetic beta particles or

Table 8.1 Air Monitoring by Task

Monitoring Equipment: Specify by task. Indicate type as necessary. Attach calibration sheets/graphs and manufacturer's instructions.

Instrument	Task	Action Guidelines	Comments
O₂/Combustible Gas Indicator (CGI)	1 2 3 4 5	0–10% LEL — No explosion hazard. Proceed w/caution, continuous monitoring 10% — Explosion hazard; interrupt task/evacuate, reassess >10% LEL 21.0% O₂ — Oxygen normal <19.5% O₂ — Oxygen deficient; notify SSHO. >22.5% O₂ — Interrupt task/evacuate	Not used.
Photoionization Detector Lamp Type ( ) 11.7 eV (X) 10.2 eV ( ) 9.8 eV ( ) [r/3] eV	1 2 3 4 5	Specify: PID	Initial site entry and throughout excavation when personnel approach the excavation site or engage in sampling within the excavation area. Repeat of sampling after successive zero readings will be at the discretion of the SSHO; however, whenever additional stained soils are excavated—prior to entry into the excavation—this monitoring will be required.
Detector Tubes—Colorimetric Type Benzene Type	1 2 3 4 5	Specify:	Not used.
Personnel Monitor—Low Volume Type Type	1 2 3 4 5	Specify: For initial site characterization and in the event that visible dust is present, monitor using a low-volume air-monitoring pump set at 2 l/min. Monitor for at least 4 h and send cassette to laboratory for analysis.	See immediately preceding pages for discussion of sampling required.
Other:	1 2 3 4 5		

gamma rays). Periodically the film in the badge is replaced, and the exposed film is for-warded to a laboratory for analysis. The density of the developed film is proportional to the exposure received. The various filters reveal the type and energy of the radiation. Thus the badge report can indicate deep exposure that can be construed as whole body exposure or shallow exposure that represents skin exposure. These values are measured to insure that exposures are below those listed in Table 8.1.

Another commonly used dosimeter is the thermo-luminescent dosimeter (TLD). The TLD consists of a small chip of material (e.g., LiF or CaF_2) that, when heated after an exposure to penetrating radiation, gives off light in proportion to the exposure received. TLDs are commonly found in badges with filters similar to film badges and are often used within rings worn by individuals handling relatively large quantities of energetic beta- or gamma-emitting radionuclides (e.g., ^{32}P, ^{137}Cs). Ring badges are used to determine the dose to extremities.

By examining monthly exposure reports, trends in exposures or higher than usual exposures may indicate that there is a problem with contamination or radiation safety pro-cedures.

Assessing internal radiation exposures is far more difficult than determining external exposure. Procedures with this purpose are collectively termed *bioassays*. For many water soluble compounds containing low-energy beta emitters (e.g., ^3H, ^{14}C), the bioassay con-sists of a urinalysis utilizing liquid scintillation counting. For radioiodine internal exposure may be assessed by using a NaI scintillation probe to externally measure the amount of radiation coming from the thyroid.

8.2 EVACUATION ZONES

Air monitoring is one of the tools used to determine the location of evacuation zones. Evacuation zones are used to provide safe refuge for on-site personnel and the approach-ing public in emergency contaminant release situations. Air monitoring is also used to determine the effect of contaminant releases on the surrounding environment. Emergency response plans must include air-monitoring protocols for area and perimeter monitoring whenever the potential for area and off-site contaminant dispersion is present.

All evacuation routes will be designated to move personnel away from a hazardous area in a safe and efficient manner and to establish efficient traffic patterns for fire and emergency equipment during an emergency response. These evacuation routes will be located at a safe distance upwind of all areas of activities. Personnel accounting should be a requirement at each emergency evacuation assembly point.

8.2.1 Emergency Equipment Locations

The anticipated dispersion pathways of site or facility contaminant also determine the location of emergency equipment on-site. Emergency equipment should be stored near hazardous areas; however, not so near that during an incident, approach cannot be made to don emergency equipment. Multiple storage locations out of the anticipated path of con-taminants may be necessary to access emergency equipment.

Safety and emergency equipment should include the following:

- For rescue purposes, two positive pressure self-contained breathing apparatus (SCBA) units dedicated and marked "for emergency only"

- Emergency eyewashes and showers in compliance with ANSIZ Z358.1
- Fire extinguishers with a minimum rating of 20-A: 120-B: C, or as appropriate to the chemical hazard (The use of fire extinguishers and fire suppressions systems may influence air-monitoring protocol changes.)

Emergency equipment containing neoprene seals may fail in an atmospheric emergency when ammonia or high concentrations of volatiles are present. Self Contained Breathing Apparatus (SCBA) regulator valves have neoprene seals and, thus, SCBAs used on certain atmospheres may fail. Use air monitoring to determine whether approach or sustained presence in a chemical risk area can be maintained.

8.2.2 Site Security and Control

In cases where an emergency situation does not pose a threat to the public and off-site emergency response teams are not dispatched to the site, a responsible on-site party must coordinate the appropriate emergency response and communicate with the public as necessary.

However, if an emergency arises that presents an immediate threat to the public or otherwise requires additional support, the emergency response system for the site or facility should be activated in the manner prescribed by the off-site emergency response organization. This response should include air monitoring to determine the extent of off-site risk and to establish site zones.

Emergency response teams at hazardous waste sites are led by an incident commander. Emergency response at other chemical or radioactive sites may also be led by an incident commander. All air-monitoring results should be made available to the incident commander.

8.2.3 Incident/Accident Report

Reports of incidents/accidents should include the following:

- Name and telephone number of reporter
- Name and address of facility
- Time and type of incident (e.g., release, fire)
- Name and quantity of material(s) involved, to the extent known, and the location of the discharge within the facility
- The extent of injuries
- The possible hazards to human health, or the environment, outside the site area
- Actions the person reporting the discharge proposes to take to contain, clean up, and remove the substance

For sites with airborne hazard potential, air-monitoring information must be included in this report to substantiate the hazard analysis and provide information on personnel exposures. Area and perimeter air-monitoring results must also be attached.

All real-time air-monitoring results that could influence needed medical treatment and decisions at emergency rooms must be provided to the medical staff. The air-monitoring results, both real time and laboratory analytical, should be made part of the employee personnel records and also provided to the medical staff after an environmental incident.

8.3 SITE WORK ZONE

Sample for breathing zone (BZ) concentrations of contaminants to establish respiratory and other PPE requirements. All exposures are calculated without regard to respiratory protection.

8.3.1 Integrated Sampling Example

Collect full shift (for at least 7 continuous hours) personal samples, including at least one sample for jobs classified as worst-case scenarios of the worker's regular, daily exposure to lead. An air-sampling pump will be worn by the individual with the highest potential for exposure. A filter cassette attached to the pump will be used to collect particulates for later analysis to determine particulate exposure during on-site work. The TWA will be calculated using the "real time," not defaulting to a value of 8 h (e.g., if workers wear the pump 3 h, the TWA will be for that 3 h, not for 8 h with assumed nonexposure for the other 5 h).

- Each day before use, perform a leak test on the pumps according to the manufacturer's instructions.
- Calibrate each personal sampling pump with a representative sampler in line.
- Use the sample and analysis procedures prescribed in the NIOSH 7105 or 7300 method for the lead particulate samples collected using the air-sampling pumps.
- Sample at an accurately known flow rate between 1 and 4 l/min.
- Do not exceed a filter loading of 2 mg total dust.
- Take readings in the BZ of the employee expected to have the greatest exposure potential.

A Mini-Ram will be used in addition to integrated air sampling to monitor exposures. The Mini-Ram is a real-time instrument. Action levels (ALs) will be based on the adjusted exposure limits (AELs). When the AELs exceed the AL for lead (0.03 mg/m^3), respiratory protection will be required. The AELs are calculated as follows:

$$AELs = [(1 \times 10^6 \text{ mg/kg})(0.03 \text{ mg/m}^3)]/\text{soil concentration}$$

The "worst-case" soil concentration of lead is 380,000 mg/kg. Using this value, when the Mini-Ram reading is 0.08, respiratory protection will be required. When the Mini-Ram reading is 4.0, back off the site.

If worker exposure data based on air-monitoring measurements confirm that no employee is exposed to airborne lead concentrations at or above the AL, make a written record of this determination. This record will include as a minimum the date of determination, location within the worksite, and the name and social security number of each employee monitored.

Additional personal monitoring will be required if an employee develops symptoms indicating possible exposure to hazardous substances or if increased sampling frequency is required by the site's air-monitoring professional.

8.3.2 Field QA and QC Example

Implement the following controls to ensure monitoring is accurate, reliable, and representative of the probable worst conditions:

- Monitor employees with the highest expected exposures.
- Ensure air sample analyses are performed by a laboratory that has been judged proficient in four successive round robins of the AIHA PAT program.
- Together with sample results, keep records on laboratory procedures, including analyses of sealed field and lab blanks, equipment checks and calibration, and notations on problems that may have affected the sample results.

8.3.3 Invasive Work Sampling Example

Oxygen, explosive atmospheres (methane), and toxic substances (benzene, hydrogen sulfide, and vinyl chloride) will be monitored to determine respirator, engineering control, and ventilation requirements. All workers will initially wear HEPA cartridge-equipped negative air pressure respirators.

- Test for oxygen, flammable gases, and hydrogen sulfide using a calibrated O_2/CGIs equipped with additional toxin sensors for hydrogen sulfide. The LEL readout will be used as an indication of the presence of flammable gases, including methane.
- If testing indicates the presence of less than 19.5% oxygen, more than 10% LEL, or more than 5 ppm hydrogen sulfide, back off and ventilate the space until testing shows the levels are within permissible limits.

To detect if any chemicals are being volatilized, a PID PI-101 will be used to scan the sampling sites. If methane has been detected using the CGI, PID readings will be suspect.

In the event that the PID displays a sustained deflection of 1 ppm (defined as needle deflection that indicates a reading of 1 ppm or 1 ppm above background for 1 min without intervening zero readings) or any reading above 5 ppm, sampling will cease, and all on-site workers will don organic vapor cartridges in addition to HEPA cartridges. (Respirators must have stacked cartridge holders, or combination HEPA-organic vapor cartridges must be available.)

- Benzene and vinyl chloride colorimetric Sensidyne detector tubes are used to indicate if either chemical is being volatilized.
- If the benzene detector tube indicates the benzene concentration is greater than 0.5 ppm, but less than 20 ppm, continue work using HEPA-organic vapor-equipped negative air pressure respirators.
- If the vinyl chloride detector tube indicates vinyl chloride is present in concentrations greater than 1 ppm, back off. Cartridge-equipped negative air pressure respirators are not available for vinyl chloride and other volatiles in combination, thus, negative air pressure respirators will not be used when vinyl chloride is detected above 1 ppm.

8.3.4 Sampling and Initial Site Work Hazard Analysis Example

8.3.4.1 Perimeter Monitoring

The site boundaries clearly mark off the "clean" off-site areas from the "contaminated" on-site areas; chemical contamination from the site should not be a hazard associated with perimeter and off-site monitoring.

Site Walk-Through, Site Surveys, Sample Grid Layout

General hazards associated with site walk-through, site surveys, and sampling grid layout include the following:

- Exposure to irritant and toxic plants such as poison ivy and sticker bushes may cause allergic reactions.
- Surfaces covered with heavy vegetation and undergrowth create a tripping hazard.
- Back strain may be due to carrying instruments.
- Native wildlife such as rodents, ticks, and snakes present the possibility of bites; many animals and insects are disease vectors for diseases such as Lyme disease.
- Driving vehicles on uneven or unsafe surfaces can result in accidents such as overturned vehicles or flat tires.
- Avoid heat stress/cold stress exposure.
- Avoid on-site chemical hazards depending on contaminant location and contact or disturbances of contaminated areas.

Hazard Prevention

- Wear long-sleeved disposable clothing to minimize contact with irritant and toxic plants and to protect against insect bites.
- Render appropriate first aid for an individual's known allergic reactions.
- Step carefully to avoid terrain hazards and to minimize slips and falls. Steel-toed boots provide additional support and stability.
- Use proper lifting techniques to prevent back strain.
- Avoid wildlife when possible. In case of an animal bite, perform first aid and capture the animal, if possible, for rabies testing.
- Check for ticks after leaving a wooded or vegetated area.
- A site surveillance on foot might be necessary to choose clear driving paths. Vehicles are prohibited on the site with the exception of the drill rig equipment.
- Implement heat stress management techniques such as shifting work hours, fluid intake, and monitoring employees for symptoms, especially high-risk workers.

8.3.4.2 Air Sampling and Monitoring Example

General hazards frequently encountered during air sampling and monitoring include the following:

- Hazards associated with the sampling the ambient environment.
- Readings indicating nonexplosive atmospheres, low concentrations of toxic substances, or other conditions may increase or decrease suddenly, changing the associated risks

Hazard Prevention

- Familiarize workers with the use, limitations, and operating characteristics of the monitoring instruments.
- Use only intrinsically safe equipment.
- Perform continuous monitoring in variable atmospheres.
- Use intrinsically safe instruments.

8.3.4.3 Water Sampling Example

Both physical and chemical hazards are associated with water sampling, and they include contact with contaminated water.

Hazard Prevention

- The buddy system must be used at all times.
- Use chemical resistant clothing.

8.3.4.4 Surface Soil/Sediment Sampling Example

For the purposes of this hazard identification section, surface soil/sediment sampling will be considered for any soil sampling completed by hand using a trowel, split spoon, shovel, auger, or other type of handheld tool. Hazards generally associated with soil and tailings/spoils sampling include the following:

- Contact with or inhalation of contaminants, potentially in high concentrations in sampling media.
- Back strain and muscle fatigue due to lifting, shoveling, and augering techniques.
- Contact with or inhalation of decontamination solutions.

Hazard Prevention

To minimize exposure to chemical contaminants, a thorough review of suspected contaminants must be completed and an adequate protection program implemented.

- Proper lifting (prelift weight assessment, use of legs, multiple personnel) techniques will prevent back strain. Use slow easy motions when shoveling, augering, and digging to decrease muscle strain.
- **Note:** The surface soils will be disturbed. In order to guard against dust generation, any dry soils will be wetted down with a light mist. The mist will be applied with handheld low-pressure misting bottles or a fire hose equipped with a mist nozzle, whichever is most efficient. Thus inhalation of dust particulates and the chemicals of concern potentially absorbed to these particles should not be a primary exposure pathway for workers 6–8 ft from the sampling sites. However, if during the site activities, visible dust is apparent due to windy conditions or lack of effective wetting, further wetting of the work area surface is necessary.

8.4 RADIATION SITES

Radionuclides in various chemical and physical forms have become extremely important tools in modern business, industry, research, and teaching. Radioactive materials are incorporated in many manufactured goods and are used in many industrial services. The use of radioactive materials generates radioactive waste. The ionizing radiations emitted by these materials and wastes, however, can pose a hazard to human health. For this reason special handling precautions must be observed with radionuclides.

8.4.1 Atomic Structure

The atom, which has been referred to as the "fundamental building block of matter," is itself composed of three primary particles: the proton, the neutron, and the electron. Protons and neutrons are relatively massive compared to electrons and occupy the dense core of the atom known as the nucleus. Protons are positively charged, while neutrons, as their name implies, are neutral. The negatively charged electrons are found in an extended cloud surrounding the nucleus.

The number of protons within the nucleus defines the atomic number, designated by the symbol Z. In an electrically neutral atom (i.e., one with equal numbers of protons and electrons), Z also indicates the number of electrons within the atom. The number of protons plus neutrons in the nucleus is termed the atomic mass, symbol A. For lighter elements the number of neutrons in a stable nucleus approximately equals the number of protons.

The atomic number of an atom designates its specific elemental identity. For example, an atom with a Z = 1 is hydrogen, an atom with Z = 2 is helium, while Z = 3 identifies an atom of lithium. Atoms characterized by a particular atomic number and atomic mass are called nuclides. A specific nuclide is represented by its chemical symbol with the atomic mass in a superscript (e.g., ^{3}H, ^{14}C, ^{125}I). Nuclides with the same number of protons (i.e., same Z) but different number of neutrons (i.e., different A) are called isotopes. Isotopes of a particular element have nearly identical chemical properties.

8.4.2 Radioactive Decay

Depending on the ratio of neutrons to protons within its nucleus, an isotope of a particular element may be stable or unstable. Over time, the nuclei of unstable isotopes spontaneously disintegrate or transform in a process known as radioactive decay or radioactivity. As part of this process, various types of ionizing radiation may be emitted from the nucleus. Nuclides that undergo radioactive decay are called radionuclides. This is a general term as opposed to the term *radioisotope* that is used to describe a particular relationship. For example, ^{3}H, ^{14}C, and ^{125}I are radionuclides. Tritium (^{3}H), on the other hand, is a radioisotope of hydrogen.

Some radionuclides such as ^{14}C, ^{40}P, and ^{238}U occur naturally in the environment, while others such as ^{32}Ph or ^{32}Na are produced in nuclear reactors or particle accelerators. Any material that contains measurable amounts of one or more radionuclides is referred to as a radioactive material.

8.4.3 Activity

The quantity that expresses the degree of radioactivity or radiation-producing potential of a given amount of radioactive material is activity. The most commonly used unit of activity is the curie (Ci), which was originally defined as that amount of any radioactive material that disintegrates at the same rate as 1 g of pure radium, which equals 3.7×10^{10} disintegrations per second (dps). A millicurie (mCi) = 3.7×10^7 dps (2.22×10^9 dpm) and a microcurie (μCi). = 3.7×10^4 dps (2.22×10^6 dpm). The activity of a given amount of radioactive material is independent of the mass of the element present and is determined only by the disintegration rate. Thus, two 1-Ci sources of ^{137}Cs might have very different masses depending on the relative proportion of nonradioactive atoms present in each source.

8.4.4 Decay Law

The rate at which a quantity of radioactive material decays is proportional to the number of radioactive atoms present. This quantity can be expressed by the equation

$$dN/dt = \lambda N \tag{1}$$

where dN/dt is the disintegration rate of the radioactive atoms, λ is the decay constant, and N is the number of radioactive atoms present at time t. Integration of this equation and expressing it in exponential form yields:

$$N = N_o e^{-\lambda t} \tag{2}$$

where N_o is the initial number of radioactive atoms present and e is the base of the natural logarithms. Because activity A is proportional to N, the equation is often expressed as

$$A = A_o e^{-\lambda t} \tag{3}$$

8.4.5 Half-Life

As N decreases over time, dN/dt decreases proportionately. For example, when half of the radioactive atoms in a given quantity of radioactive material have decayed, the disintegration rate (or activity) is also halved. The time required for the activity of a quantity of a particular radionuclide to decrease to half its original value is called the half-life ($T_{1/2}$) for the radionuclide. Table 8.2 indicates half-lives and other characteristics of several radionuclides used in research.

It can be shown mathematically that the $T_{1/2}$ of a particular radionuclide is related to λ as follows:

$$\lambda = \frac{\ln 2}{T_{1/2}} = \frac{0.693}{T_{1/2}} \tag{4}$$

Substituting this value of λ into Equation 3, one gets:

$$A = A_o e^{\frac{-0.693t}{T_{1/2}}} \tag{5}$$

174

Table 8.2   Characteristics of Selected Radionuclides

Radionuclide	Half-Life	Type & Max. Energy (MeV)
^{3}H	12.3 years	0.0186
^{14}C	5370 years	0.155
^{35}S	87.2 days	0.167
^{45}Ca	163 days	0.252
^{51}Cr	27.7 days/X	0.320
^{12}I	559.7 days/X	0.035
^{13}I	18.0 days/X	0.606/0.364
^{32}P	14.3 days	1.71
^{99}Tc	6.0 hours/X	0.140/0.142

Example 1: A researcher obtains 5 mCi of ^{32}Ph ($T_{1/2}$ = 14.3 days). How much activity will remain after ten days?

$$A = ?$$

$$A_o = 5 \text{ mCi}$$

$$t = 10 \text{ days}$$

$$\lambda = 0.693/14.3$$

$$A = A_o e^{-\lambda t} = 5e^{[-0.693(10)]/14.3} = 3.1 \text{ mCi}$$

An alternative method of determining the activity of a radionuclide remaining after a given time is through the use of

$$f = 1/2^n \tag{6}$$

where f equals the fraction of the initial activity remaining after time t and n equals the numbers of half-lives that have elapsed. In Example 1 above,

$$n = t/T_{1/2}$$

$$n = 10/14.3 = 0.69$$

$$f = 1/2^{0.69} = 0.62$$

$$A = fA_o = (0.62)(5) = 3.1 \text{ mCi}$$

The remaining fraction f for a given time n may be found in the literature from the manufacturer enclosed with most short-lived radionuclides. Both methods may be used to calculate activities at a prior date t and thus may be negative.

8.4.6 Types of Ionizing Radiation

Ionizing radiation may be electromagnetic or may consist of high-speed subatomic particles of various masses and charges.

8.4.6.1 Alpha Particles

Certain radionuclides of high atomic mass (e.g., ^{226}Ra, ^{238}U, ^{239}Pu) decay by the emission of alpha particles. These are tightly bound units of two neutrons and two protons each (a helium nucleus). Emission of an alpha particle results in a decrease of two units of atomic number (Z) and four units of atomic mass (A). Alpha particles are emitted with discrete energies characteristic of the particular transformation from which they originate.

8.4.6.2 Beta Particles

A nucleus with a slightly unstable ratio of neutrons to protons may decay through the emission of a high-speed electron called a beta particle. This emission results in a net increase of one unit of atomic number (Z). The beta particles emitted by a specific radionuclide range in energy from near zero up to a maximum value characteristic of the particular transformation.

8.4.6.3 Gamma Rays

A nucleus that is in an excited state may emit one or more photons (i.e., particles of electromagnetic radiation) of discrete energies. The emission of these gamma rays does not alter the number of protons or neutrons in the nucleus, but instead has the effect of moving the nucleus from a higher to a lower energy state. Gamma-ray emission frequently follows beta decay, alpha decay, and other nuclear decay processes.

X-rays and gamma rays are electromagnetic radiation, as is visible light. The frequencies of X-rays and gamma rays are much higher than that of visible light and so each carries much more energy. Gamma and X-rays cannot be completely shielded. They can be attenuated by shielding, but not stopped completely. A gamma-emitting nuclide may yield multiple gamma rays and X-rays, each with its own discrete energy. It is possible to identify a gamma-emitting nuclide by its spectrum.

8.4.6.4 X-rays

X-rays are also part of the electromagnetic spectrum and are distinguished from gamma rays only by their source (i.e., orbital electrons, rather than the nucleus). X-rays are emitted with discrete energies by electrons as they shift orbits following certain types of nuclear decay processes.

8.4.7 Rules of Thumb

The activity of any radionuclide is reduced to less than 1% after 7 half-lives and less than 0.1% after 10 half-lives (i.e., $2^{-7} \times 100 = 0.8\%$ and $2^{-10} \times 100 = 0.09\%$) (Table 8.3).

8.4.8 Excitation/Ionization

The various types of radiation (e.g., alpha particles, beta particles, and gamma rays) *impart their energy to matter* primarily through excitation and ionization of orbital electrons. The term *excitation* is used to describe an interaction where electrons acquire energy

Table 8.3

Radiation	Energy (keV)	Decay %
Gamma	35	6.7
Ka X-ray	27.4	114
Kb X-ray	31	25.6
L X-ray	3.9	12
K Conv. Elec.	3.7	80
L Conv. Elec.	31	11.8
M + Conv. Elec.	35	2.5
K Auger Elec.	23	20
L Auger Elec.	3–4	160

from a passing charged particle, but are not removed completely from their atom. Excited electrons may subsequently emit energy in the form of X-rays during the process of returning to a lower energy state. The term *ionization* refers to the complete removal of an electron from an atom following the transfer of energy from a passing charged particle. Any type of radiation having sufficient energy to cause ionization is referred to as ionizing radiation. In describing the intensity of ionization, the term *specific ionization* is often used. Specific ionization is defined as the number of ion pairs formed per unit path length for a given type of radiation.

8.4.9 Characteristics of Different Types of Ionizing Radiation

Alpha particles have a high specific ionization and a relatively short range. Alpha particles travel in air only a few centimeters, while in tissue they travel only fractions of a millimeter. For example, an alpha particle cannot penetrate the dead cell layer of human skin.

Beta particles have a much lower specific ionization than alpha particles and a considerably longer range. The relatively energetic betas from ^{32}P have a range of 6 m in air or 8 mm in tissue. Only 6 mm of air or 5 μm of tissue, on the other hand, stop the low-energy betas from ^{3}H.

Gamma and X-rays are referred to as indirect ionizing radiation because, having no charge, they do not directly apply impulses to orbital electrons as do alpha and beta particles. A gamma ray or X-ray instead proceeds through matter until it undergoes a chance interaction with a particle. If the particle is an electron, it may receive enough energy to be ionized, whereupon it causes further ionization by direct interactions with other electrons. The net result is that indirectly ionizing particles liberate directly ionizing particles deep inside a medium, much deeper than the directly ionizing particles could reach from the outside. Because gamma rays and X-rays undergo only chance encounters with matter, they do not have a finite range. In other words a given gamma ray has a definite probability of passing through any medium of any depth.

8.4.10 Exposure (roentgen)

Exposure is a measure of the strength of a radiation field at some point. It is usually defined as the amount of charge (i.e., sum of all ions of one sign) produced in a unit mass of air when the interacting photons are completely absorbed in that mass. The most commonly used unit of exposure is the roentgen (R), which is defined as that amount of X or gamma radiation that produces 2.58×10^{-4} C/kg of dry air. In cases where exposure is to be expressed as a rate, the unit would be roentgens per hour or more commonly, milliroentgens per hour. A roentgen refers to the ability of *photons* to ionize *air*. Roentgens are very limited in their use. They apply only to photons, only in air, and only with an energy under 3 MeV. Because of their limited use, no new unit in the SI system has been chosen to replace it.

8.4.11 Absorbed Dose (rad)

Whereas exposure is defined for air, the absorbed dose is the amount of energy imparted by radiation to a given mass of any material. The most common unit of absorbed dose is the rad (radiation absorbed dose), which is defined as a dose of 100 ergs of energy per gram of the material in question. Absorbed dose may also be expressed as a rate with units of rads per hour or millirads per hour.

$$\text{New SI Unit: } 1 \text{ gray (Gy)} = 1 \text{ J/kg} (= 100 \text{ rads}) \tag{7}$$

8.4.12 Dose Equivalent (rem)

Although the biological effects of radiation are dependent on the absorbed dose, some types of particles produce greater effects than others for the same amount of energy imparted. For example, for equal absorbed doses, alpha particles may be ten times as damaging as beta particles. To account for these variations when describing human health risk from radiation exposure, the quantity dose equivalent is used. This quantity is the absorbed dose multiplied by certain "quality" and "modifying" factors indicative of the relative biological damage potential of the particular type of radiation.

The unit of dose equivalent is the rem (roentgen equivalent in man) or, more commonly, millirem. For gamma ray or X-ray exposures, the numerical value of the rem is essentially equal to that of the rad.

$$\text{New SI Unit: } 1 \text{ Sievert (Sv)} = 1 \text{ Gy} \times Q \tag{8}$$

Some quality factors are listed below. (Note that there is quite a bit of discrepancy between different agencies' values (Table 8.4).

Table 8.4

Radiation Type	NRC	ICRU	NCRP
X-rays & gamma rays	1	1	1
Beta rays except ^{3}H	1	1	1
Tritium betas	1		1
Fast neutrons	10	25	20
Alpha particles	20	25	20

Table 8.5   Sample Weighting Factors

Gonads	0.25
Breast	0.15
Lung	0.12
Bone	0.03
Marrow	0.12
Remainder	0.30

8.4.13 Effective Dose Equivalent

Modifying the dose equivalent by a weighting factor that relates to the radiosensitivity of each organ and summing these weighted dose equivalents produces the effective dose equivalent (see Table 8.5). (A complete list of applications and procedures is beyond the scope of this guide.)

8.4.14 Biological Effects of Ionizing Radiation

The energy deposited by ionizing radiation as it interacts with matter may result in the breaking of chemical bonds. If the irradiated matter is living tissue, such chemical changes may result in an altered structure of function of constituent cells.

Because the cell is composed mostly of water, less than 20% of the energy deposited by ionizing radiation is absorbed directly by macromolecules. More than 80% of the energy deposited in the cell is absorbed by water molecules, with the resultant formation of highly reactive free radicals.

These radicals and their products (e.g., hydrogen peroxide) may initiate numerous chemical reactions that result in damage to macromolecules and a corresponding alteration of structure or function. Damage produced within a cell by the radiation-induced formation of free radicals is described as the indirect action of radiation.

As a result of the chemical changes in the cell caused by the direct or indirect action of ionizing radiation, large biological molecules may undergo a variety of structural changes that lead to altered function. Some of the more common effects that have been observed are inhibition of cell division, denaturation of proteins and inactivation of enzymes, alteration of membrane permeability, and chromosome aberrations.

8.4.14.1 Radiosensitivity

The cell nucleus is the major site of radiation damage that leads to cell death. DNA within the nucleus controls all cellular function. Damage to the DNA molecule may prevent it from providing the proper template for the production of additional DNA or RNA. This hypothesis is supported by research that has shown that cells are most sensitive to radiation damage during reproductive phases (i.e., during DNA replication).

In general, it has been found that cell radiosensitivity is directly proportional to reproductive capacity and inversely proportional to the degree of cell differentiation. The following list of cells illustrates this general principle:

Very radiosensitive:

• Vegetative intermitotic cells

- Mature lymphocytes
- Erythroblasts and spermatogonia
- Basal cells
- Endothelial cells

Moderately radiosensitive:

- Blood vessels and interconnective tissue
- Osteoblasts
- Granulocytes and osteocytes
- Sperm erythrocytes

Relatively radioresistant:

- Fixed postmitotic cells
- Fibrocytes
- Chondrocytes
- Muscle and nerve cells

The considerable variation in the radiosensitivities of various tissues is due, in part, to the differences in the sensitivities of the cells that compose the tissues. Also important in determining tissue sensitivity are such factors as the state of nourishment of the cells, interactions between various cell types within the tissue, and the ability of the tissue to repair itself.

The relatively high radiosensitivity of tissues consisting of undifferentiated, rapidly dividing cells suggest that, at the level of the human organism, a greater potential exists for damage to the fetus or young child than to an adult for a given dose. This tendency has, in fact, been observed in the form of increased birth defects following irradiation of the fetus and an increased incidence of certain cancers in individuals who were irradiated as children.

8.4.15 Human Health Effects

The effects of ionizing radiation described at the level of the human organism can be divided broadly into one of two categories: stochastic or nonstochastic effects.

8.4.15.1 Stochastic Effects

As implied from the name, "stochastic" effects occur by chance. Stochastic effects caused by ionizing radiation consist primarily of genetic defects and cancer. As the dose to an individual increases, the probability that cancer or a genetic defect will occur also increases. However, at no time, even for high doses, is it certain that cancer or genetic damage will result. Similarly, for stochastic effects, there is no threshold dose below which it is relatively certain that an adverse effect cannot occur. Because stochastic effects can occur in unexposed individuals, one can never be certain that the occurrence of cancer or genetic damage in an exposed individual is due to radiation.

8.4.15.2 Nonstochastic Effects

Unlike stochastic effects, nonstochastic effects are characterized by a threshold dose below which they do not occur. The magnitude of the effect is directly proportional to the size of the dose. Furthermore, for nonstochastic effects there is a clear causal relationship between radiation exposure and the effect. Examples of nonstochastic effects include sterility, erythema (skin reddening), and cataract formation. Each of these effects differs from the other in both its threshold dose and in the time over which this dose must be received to cause the effect (i.e., acute vs. chronic exposure).

The range of nonstochastic effects resulting from an acute exposure to radiation is collectively termed "radiation syndrome." This syndrome may be subdivided as follows:

- Hemopoietic syndrome—characterized by depression or destruction of bone marrow activity with resultant anemia and susceptibility to infection (whole body dose of about 200 rads)
- Gastrointestinal syndrome—characterized by destruction of the intestinal epithelium with resultant nausea, vomiting, and diarrhea (whole body dose of about 1000 rads)
- Central nervous system syndrome—direct damage to nervous system with loss of consciousness within minutes (whole body doses in excess of 2000 rads). The LD_{50} (i.e., dose that would cause death in half of the exposed population) for acute whole body exposure to radiation in humans is about 450 rads.

8.4.16 Determinants of Dose

The effect of ionizing radiation upon humans or other organisms is directly dependent on the size of the dose received (although dose rate may also be important). The dose, in turn, is dependent on a number of factors, including the strength of the source, the distance from the source to the affected tissue, and the time over which the tissue is irradiated. The manner in which these factors operate to determine the dose from a given exposure differs significantly for exposures that are "external" (i.e., resulting from a radiation source located outside the body) and those that are "internal" (i.e., resulting from a radiation source located within the body).

8.4.16.1 External Exposures

Exposures to sources of radiation located outside the body are of concern primarily for sources emitting gamma rays, X-rays, or high-energy beta particles. External exposures from radioactive sources that emit alpha or low-to-medium energy beta particles are not significant because these radiations do not penetrate the dead outer cell layer of the skin.

As with all radiation exposures, the size of the dose resulting from an external exposure is a function of

- The strength of the source
- The distance from the source to the tissue being irradiated
- The duration of the exposure

In contrast to the situation for internal exposures, however, these factors can be altered (either intentionally or inadvertently) for a particular external exposure situation, with a resultant modification of the dose received.

The effectiveness of a given dose of external radiation in causing biological damage is dependent on the portion of the body irradiated. For example, because of differences in the radiosensitivity of constituent tissues, the hand is far less likely to suffer biological damage from a given dose of radiation than are the gonads. Similarly, a given dose to the whole body has a greater potential for causing adverse health effects than does the same dose to only a portion of the body.

8.4.16.2 Internal Exposures

Exposures to ionizing radiation from sources located within the body create cause for concern. Of particular concern are internally emitted alpha and beta particles that cause significant damage to tissue when they deposit their energy along highly localized paths.

In contrast to the situation for external exposures, the source to tissue distance, exposure duration, and source strength cannot be altered for internal radiation sources. Instead, once a quantity of radioactive material is taken up by the body (e.g., by inhalation, ingestion, or absorption), an individual is "committed" to the dose that will result from the quantities of the particular radionuclide(s) involved.

In general, radionuclides taken up by the body do not distribute equally throughout the body's tissues. Often, a radionuclide concentrates in a critical organ (defined as the organ for which damage from uptake of the particular radionuclide leads to the greatest damage to the body as a whole). For example, ^{131}I and ^{125}I concentrate in the thyroid, ^{45}Ca and ^{32}P in the bone, and ^{59}Fe in the spleen.

The dose committed to a particular organ or portion of the body depends, in part, on the time over which these areas of the body are irradiated by the radionuclide. This dose, in turn, is determined by the radionuclide's physical and biological half-lives (i.e., the effective half-life). The biological half-life of a radionuclide is defined as the time required for one half of a given amount of radionuclide to be removed from the body by normal biological turnover.

8.4.17 Sources of Exposure

Exposures to ionizing radiation can be classified broadly according to whether they result from sources within the work environment (i.e., occupational exposures) or from sources outside the work environment (i.e., nonoccupational exposures).

8.4.17.1 Occupational Exposure

Occupational exposures are those received by individuals as a result of working with or near radiation sources (i.e., radioactive material or radiation-producing devices). Occupational exposures differ from nonoccupational exposures in that they are generally received during the course of a 40-hour work week (exposures from natural background are received continuously over 168 hours per week). Whereas the nonoccupational exposure received by a given individual is largely unknown, an individual's occupational exposure is closely monitored and controlled.

8.4.17.2 Nonoccupational Exposure

Nonoccupational exposures can be divided into one of two categories: those originating from natural sources or those resulting from man-made sources.

All individuals are continuously exposed to ionizing radiation from various natural sources. These sources include cosmic radiation and naturally occurring radionuclides within the environment and within the human body. The radiation levels resulting from natural sources are collectively referred to as "natural background." Natural background and the associated dose it imparts varies considerably from one location to another in the U.S. It is estimated that the average whole body dose equivalent from natural background in the U.S. is about 250 mrem/person/year.

The primary source of man-made nonoccupational exposures is medical irradiation, particularly diagnostic procedures (e.g., X-ray and nuclear medicine examinations). Such procedures, on average, contribute an additional 100 mrem/person/year in the U.S. All other sources of man-made, nonoccupational exposures, such as nuclear weapon fallout, nuclear power plant operations, and the use of radiation sources in industry and universities, contribute an average of less than 1 mrem/person/year in the U.S.

8.4.18 Exposure Limits

Concern over the biological effects of ionizing radiation began shortly after the discovery of X-rays in 1895. From that time to the present, numerous recommendations regarding occupational exposure limits have been proposed and modified by various radiation protection groups, the most important being the International Commission on Radiological Protection (ICRP). These guidelines have, in turn, been incorporated into regulatory requirements for controlling the use of materials and devices emitting ionizing radiation.

8.4.19 Basis of Recent Guidelines

The guidelines established for radiation exposure generally have two principal objectives:

1. The prevention of acute radiation effects (e.g., erythema, sterility)
2. The limiting of the risks of late, stochastic effects (e.g., cancer, genetic damage) to "acceptable" levels

Numerous revisions of standards and guidelines have been made over the years to reflect both changes in the understanding of the risk associated with various levels of exposure and changes in the perception of what constitutes an "acceptable" level of risk.

Current guidelines for radiation exposure are based on the conservative assumption that there is no safe level of exposure. In other words even the smallest exposure has some probability of causing a late effect such as cancer or genetic damage. This assumption has led to the general philosophy of not only keeping exposures below recommended levels or regulatory limits but also of maintaining all exposures at ALARA levels. This is a fundamental tenet of current radiation safety practice.

8.4.20 Regulatory Limits for Occupational Exposure

Many of the recommendations of the ICRP and other radiation protection groups regarding radiation exposure have been incorporated into regulatory requirements by various countries. In the U.S. radiation exposure limits are found in Title 10, Part 20 of the Code of Federal Regulations (10 CFR 20).

Table 8.6   Ionizing Radiation Occupational External Exposure Limits (rems)

Systems	Quarterly	Annual
Organ systems — whole body, head and trunk, active — blood-forming organs, lens of eye — gonads	1.25	5
Hands and forearms, feet and ankles	18.75	75
Skin of the whole body	7.50	30

Table 8.6 provides a summary of the current limits for occupational external exposures.

In addition to external exposure limits, 10 CFR 20 also addresses internal exposures through the establishment of maximum permissible concentrations (MPCs) of various radionuclides allowed in air and water. The MPC is the concentration that, if breathed or ingested continuously, would result in the annual limit of intake (ALI) for that radionuclide. The ALI is, in turn, the amount of a radionuclide that, if continuously present in the body over a year, would result in the maximum allowed dose equivalent rate to the body. The values of ALIs and estimates of dose commitment per microcurie for various radionuclides are listed in Table 8.7.

8.4.21 Recommended Exposure Limits for Pregnant Workers

The increased susceptibility of the human embryo and fetus to damage from ionizing radiation presents a case for more conservative doses for pregnant workers. The National Council on Radiation Protection and Measurement (NCRP) recommends that the whole body radiation dose received by a female worker during the 9 months of her pregnancy not exceed 500 mrem (i.e., one tenth of the normal occupational dose limit). The Nuclear Regulatory Commission has published Regulatory Guide 8.13 that details potential health risks of prenatal exposures and suggests precautions and options for the pregnant worker.

The limits of exposure to members of the public from "licensed activities" are one tenth of the occupational limits. For whole body exposure this level would equal 500 mrem/year. This exposure is in addition to the 250 mrem/year received, on average, by individuals in the U.S. from natural background radiation and the 100 mrem/person/year received from medical exposures. The 500 mrem/year limit also applies to individuals under age 18 who work in the vicinity of radiation sources.

Table 8.7   ALIs for Selected Radionuclides

Radionuclide	ALI Ingested (μCi)	ALI Inhaled (μCi)
^{3}H	80,000	80,000
^{51}Cr	40,000	20,000
^{14}C	2000	2000
^{35}S	6000	20,000
^{32}P	600	900
^{137}Cs	100	200

8.4.22 Radiation Risk

The risk associated with various levels of exposure to ionizing radiation can be adequately understood only when viewed in the context of other health risks routinely encountered in life. The risk associated with various activities or situations is most often expressed as an annual mortality rate. For example, the annual mortality rate from home accidents in the U.S. is 1.1×10^{-4}. Thus, the risk of death from a home accident is approximately 1 in 10,000/year.

The American Cancer Society has reported that approximately 25% of all adults 20–65 years old will develop cancer at some time from all possible causes, such as smoking, food, alcohol, drugs, air pollutants, and natural background radiation. Thus, in any group of 10,000 workers not exposed to radiation on the job, we can expect about 2,500 to develop cancer. If this entire group of 10,000 workers were to receive an occupational radiation dose of 1000 mrem each, we could estimate that three additional cases might occur, giving a total of about 2,503. This means that a 1000 mrem dose to each of 10,000 workers might increase the cancer rate from 25% to 25.03%, an increase of about 0.03%.

As an individual, if your cumulative occupational radiation dose is 1000 mrem, your chances of developing cancer during your entire lifetime may have increased from 25% to 25.03%. If your lifetime occupational dose is 10,000 mrems, we could estimate a 25.3% chance of developing cancer. Using a simple linear model, a lifetime dose of 100,000 mrems may have increased your chances of developing cancer from 25% to 28%.

8.4.23 Philosophy of Current Radiation Safety Practice

Current regulatory limits for radiation exposure have been conservatively set to prevent all acute effects of radiation exposure and to limit the risk of chronic effects, such as cancer, to very low levels. As a primary objective, then, radiation safety practice attempts to ensure that all exposures are below these limits. The accomplishment of this objective, however, is not the ultimate aim of current radiation safety practice. The *Radiation Protection Guidance* of the Federal Agencies for Occupational Exposure, January 20, 1987, expresses the fundamental principle that:

> There shall not be any occupational exposure of workers to ionizing radiation without the expectation of an overall benefit from the activity causing the exposure.

An overriding principle of radiation protection philosophy is that all exposures must be maintained at ALARA levels. Thus, even if a given exposure to an individual is within regulatory limits, it may not be acceptable if the exposure could have been limited further by "reasonable" means.

8.4.23.1 Internal Radiation Protection

Any radionuclide, whether it be an alpha, beta, or gamma emitter, poses a potential hazard to health if it enters the body. Once within the body, the radionuclide will continue to irradiate living tissue until it is removed by natural processes. Because there is no easy way of increasing the rate of these processes, radionuclides must be prevented from being taken into the body in the first place.

To adequately protect against the uptake of radionuclides, it is important to understand the ways radionuclides enter the body. The three primary routes of uptake are as follows:

1. Ingestion
2. Inhalation
3. Absorption through the skin

8.4.23.2 Protection against Ingestion

In protecting against ingesting radioactive material, the importance of observing strict contamination control measures on-site and in the laboratory, including monitoring hands, clothing, and work areas after each procedure involving the use of radioactive material, cannot be overemphasized.

8.4.23.3 Protection against Inhalation

Certain chemical forms of particular radionuclides volatilize easily and thus pose a hazard through inhalation. Examples of these include tritiated water and radioiodine (e.g., ^{125}I, ^{131}I) in solution as NaI. Because of the volatility of these materials, procedures involving their use should always be carried out within a fume hood.

8.4.23.4 Protection against Absorption

Many of the same types of radioactive materials that pose inhalation hazards also pose a hazard by absorption through the skin. Both tritiated water and various chemical forms of radioiodine are readily absorbed through the skin. Such materials have been shown to diffuse fairly quickly through the thickness of a single plastic glove. For this reason it is important to wear two pairs of gloves and to change the outer pair frequently during procedures using these materials.

8.4.24 External Radiation Protection

Medium-to-high energy beta particles, gamma rays, and X-rays are often referred to as "penetrating" radiations because of their ability to pass through considerable thicknesses of matter. Because of this ability, penetrating radiation can originate from sources external to the body and still impart a significant dose to living tissue. Radionuclides that emit such radiations thus pose an "external" radiation hazard. Examples of such radionuclides that are used in radiotracer research are the beta emitter ^{32}P and the beta/gamma emitter ^{137}Cs.

Certain gamma emitters (e.g., ^{137}Cs, ^{60}Co) that are routinely encapsulated as sealed sources can also pose significant external hazards. On the other hand such commonly used radionuclides as ^{3}H, ^{14}C, and ^{35}S are primarily internal radiation hazards because the beta particles they emit do not have sufficient energy to penetrate the skin (although if applied directly to the skin's surface in sufficiently large quantities, ^{14}C or ^{35}S can damage living skin cells).

The dose resulting from a given exposure situation depends on the duration of the exposure, the distance from the source to the tissue, and the strength of the source. For internal exposures, these factors cannot be easily modified. However, for external exposures, the dose received can be limited by controlling these factors.

8.4.25 Minimizing Exposure Time

In general, the dose (D) received from a particular exposure situation is the product of the exposure rate (I) times the exposure time (t):

$$D = It \tag{9}$$

It follows, then, that any reduction in t spent in a radiation field (I) will decrease the total dose received.

Example 1: In order to synthesize a particular radioactive compound, a chemist must work in a radiation field of 1 mR/h. If he works 5 days per week, what is the maximum amount of time he can work in this radiation field each day and not exceed one tenth of his daily limit?

$$1/10\text{th of } (5000 \text{ mR/year})/\text{daily limit} = 250 \text{ workdays/year} \times 0.1 = 2 \text{ mR/day}$$

$$t = D/I = 2 \text{ mR/day}/1 \text{ mR/h} = 2 \text{ h/day}$$

8.4.26 Maximizing Distance from Source

If one assumes that a radiation source occupies a single point, it can be shown mathematically that I varies inversely with the square of the distance (d) from the source:

$$I_2 = (I_1) \times d_1^2/d_2^2 \tag{10}$$

where I_2 is the exposure rate at a distance of d_2 from the source and I_1 is the exposure rate at distance d_1. From this equation it is apparent that doubling the distance from a radiation source decreases the exposure rate by a factor of four. Increasing the distance by a factor of three decreases the exposure rate by a factor of nine.

Example 2: A veterinary radiologist must manipulate a ^{60}Co source with an exposure rate of 500 mR/h at 1 ft. If the source must be handled for 1 h each day, what should be the length of remote handling equipment to ensure that the individual's daily dose limit is not exceeded?

$$I_1 = 500 \text{ mR/h}$$

$$d_1 = 1 \text{ ft}$$

$$I_2 = (500 \text{ mR/year})/250 \text{ workdays/year} = 20 \text{ mR/day}$$

$$I_1 d_1^2/d_2^2 = I_2 = 500 \times 1^2/20$$

$$d_2^2 = 25 \text{ ft}^2$$

$$d_2 = 5 \text{ ft}$$

A useful adaptation for estimating the dose rate from a source of medium- to high-energy beta particles such as ^{32}P is

$$D = 200 \, A/d^2 \tag{11}$$

where D is the dose rate in millirads per hour, A is the activity of the source in curies, and d is the distance from the source in centimeters.

8.4.27 Shielding the Source

The dose from a particular exposure situation can also be reduced by decreasing the source strength (i.e., exposure rate) through the use of shielding. The amount by which a given shield reduces the exposure rate from gamma rays or X-rays is given by the following equation:

$$I = I_o e^{-\mu x}$$

where

- I_o = exposure rate at 0 absorber thickness
- x = absorber (i.e., shield) thickness
- I = exposure rate from source shielded by absorber of thickness x
- μ = attenuation coefficient (fraction of beam removed per unit thickness of absorber)

Example 3: The exposure rate from a source that emits 1 MeV gamma rays is 100 mR/h at 1 m. What will be the exposure rate if a lead brick is placed in front of the source? The attenuation coefficient (μ) for 1 MeV gamma in lead is 0.85 cm^{-1}.

$$\mu = 0.85 \text{ cm}^{-1}$$

$$x = 5 \text{ cm}$$

$$I_o = 100 \text{ mR/h}$$

$$I = ?$$

$$I = I_o e^{-\mu x} = 100e^{-(.85)(5)} = 1.4 \text{ mR/h at 1 m}$$

If instead of lead, the same thickness of Lucite (approximately the same density as water) had been utilized:

$$\mu = 0.71 \text{ cm}^{-1}$$

$$I = I_o e^{-\mu x} = 100e^{-(.071)(5)} = 70.1 \text{ mR/h at 1 m}$$

Table 8.8   Photon Attenuation Coefficients

Energy μ(cm⁻¹) (MeV)	Water	Lead	Concrete
0.01	4.99	1453	62.3
0.1	0.168	59.4	0.400
0.3	0.119	4.32	0.251
0.5	0.092	1.75	0.204
1.0	0.071	0.85	0.149
1.5	0.0057	0.59	0.121

This comparison illustrates the effectiveness of lead over Lucite in shielding gamma rays (Table 8.8).

Because beta particles have finite ranges, they do not strictly follow the shielding equation. No amount of shielding can totally absorb all gamma rays emitted by a particular source. However, for any beta-emitting source, there is a particular amount of shielding material that will absorb all beta particles emitted.

As a secondary problem, the process of beta particle absorption in matter results in the emission of X-rays known as bremsstrahlung radiation. The production of bremsstrahlung radiation is much greater for high atomic number shields (e.g., lead) than for low atomic number shields (e.g., aluminum).

The ideal material for shielding beta particles is one that is thick enough to stop the betas, but with an atomic number and density low enough to minimize bremsstrahlung production. The materials most used for this purpose are Lucite or Plexiglas, which have the added advantage of being transparent.

The range (R) of beta particles in grams per square centimeter (thickness in centimeters multiplied by the density in grams per cubic centimeter) is approximately equal to the maximum energy (E) in MeV divided by two (i.e., $R = E/2$).

8.4.28 Emergency Procedures

Despite strict adherence to all safety rules, it is likely that accidents involving radioactive material will, on occasion, occur. For this reason it is important that radioactive material handlers must know and follow the proper procedures for radiological accidents.

8.4.28.1 Personal Contamination

Personal contamination can best be removed by washing the contaminated region with warm water and soap. Do not scrub hard, use brushes, or anything that might break the skin. Remember internal contamination is much more difficult to eliminate than external contamination.

8.4.28.2 Minor Spills (Microcurie Quantities of Most Nuclides)

Notify personnel in the area that the spill has occurred. Prevent the spread of contamination by covering the spill with absorbent paper. Clean up the spill carefully using disposable gloves and remote handling tongs. Work from the outside edges in and monitor to assure adequate decontamination of an area before moving over it toward the middle. Dispose of the absorbent paper, gloves, and other contaminated materials in a plastic bag. Notify the emergency response team of the incident.

8.4.28.3 For Major Spills (Millicurie Quantities of Most Nuclides)

Clear the area. Prevent the spread of the contamination as above. Confine the movement of personnel potentially contaminated to prevent the spread. Shield the source if it can be done without further contamination or significantly increasing your radiation exposure. Close off the area to prevent entry. Notify the emergency response team immediately.

CHAPTER 9

Microcircuitry and Remote Monitoring

With the advent of microcircuitry, GIS locators, and the Internet, this chapter discusses the new monitoring options available. These protocols are particularly important in monitoring remote sites for remediation or process adequacy and safety (explosive limits and toxic chemical limit determinations).

9.1 CONTINUOUS IAQ MONITORING IN BUILDINGS

Continuous IAQ monitoring empowers IAQ professionals and facility managers to find and resolve IAQ issues quickly, before they become problems or crises. Intermittent continuous IAQ monitoring supplements the qualified professional with permanently installed scientific sensors that measure, record, and archive the indoor air quality in a building 24 h per day, 365 days per year.

One of the indicators of air quality is carbon dioxide levels within the total airstream. Carbon dioxide is generated through respiratory exhalation and incomplete combustion of fuels. Continuous monitoring systems often use infrared beams to detect carbon monoxide fluctuations (see Figure 2.6 on page 13).

Continous IAQ monitoring:

- Creates an *early warning system* for targeting and correcting IAQ issues
- Shows precisely when complaints disappear while demonstrating when most building occupants are comfortable
- Assures, through operational linkage, proper equipment operation and maintenance, even on remote sites
- Provides building owners and operators with ongoing proof of acceptable indoor air quality
- Yields a continuous historical record of contaminant levels

The early warning capability, when it is combined with multiple small unobtrusive monitoring devices, makes monitoring in sensitive areas possible. Using remote sensors, monitoring is possible within laboratories, hospitals, vehicle shops, paint booths, hazardous waste storage bunkers, and vehicular housings. In the event that the sensors detect a problem, IAQ professionals can be summoned to assess the immediate situation.

Sensors can be configured to assess multiple chemical exposure parameters. Changes in the air quality pattern would initiate a warning signal in addition to regular information transfer on-line. Further assessment with more sensitive devices may be the next requirement, or the response professional may simply document the event and issue an all-clear decision. Engineering controls, as protection from chemical overexposure, may thus be more efficient and more successful.

9.1.1 IAQ Evaluations

Building air quality has emerged as a major issue involving the health and productivity of occupants. Building-related symptoms, often referred to as sick building syndrome, have been traced to a variety of indoor air contaminants, with the underlying cause often being deficiencies in ventilation.

Air handling, conditioning, and sequencing strategies designed for ventilation systems and modification can be monitored using remote monitoring over telephone lines, dedicated cable systems, fiber optic lines, and even the Internet. In some situations proper monitoring can save energy while improving indoor air quality. An example might be the redesign of outside air handling strategies to improve the performance of an economizer cycle.

Despite the potential costs and liabilities of IAQ problems, they can often be remedied by relatively simple changes in building operation. These studies can be completed through both on-site and remote monitoring techniques.

Sorbent tube samplers are available to provide 24-h monitoring with a range of flow rates and sorbent tube types (Figure 9.1).

9.1.1.1 Characterization for IAQ Assessment

IAQ is a developing area of knowledge. The following sequential approach is used during an assessment:

- Definition with the client of the scope of work
- Serious questions of health, productivity, and credibility at stake when occupants claim that a building is making them sick

Figure 9.1 Fenceline monitoring equipment can be used for IAQ investigations, ambient air monitoring, and hazardous waste site evaluation. This model can sample a maximum of 24 sequential sorbent tube samples in a single sampling sequence. Time, date, mass flow, tube number, temperature, and wind direction and speed can be automatically logged into memory and printed or downloaded into a personal computer. (BIOS International Corp.)

- Discussion of various approaches to the investigation
- Problems often classified as either ventilation related or source related for purposes of implementing sensible corrective measures
- Coordination of efforts among team members and current building occupants
- Oversight of monitoring and diagnostic activities
- Analysis of corrective actions
- Generation of performance specifications

IAQ investigations are conducted as follows:

- Describe how to form and test explanations for the problem and its solution. The proposed approach to the building investigation may involve moving suspected contaminant sources or manipulating HVAC controls to simulate conditions at the time of complaints or to test possible corrective actions.
- IAQ measurement decision logic is developed.
- A decision to obtain IAQ-related measurements follows logically from other investigative activities.
- Before starting to take measurements, investigators need a clear understanding of how the results will be used. Without this understanding, it is impossible to plan appropriate sampling locations and times, instrumentation, and analysis procedures.
- Nonroutine measurements (such as relatively expensive sampling for VOCs are not to be conducted without site-specific justification.
- Concentrations low enough to comply with industrial occupational standards could still be harmful to children or other immune-compromised or sensitized occupants.
- Industrial IAQ problems tend to arise from high levels of individual chemical compounds, so standards set limits for individual contaminants or contaminant classes.
- Exposure workplace standards are rarely exceeded in residences and schools. Instead, IAQ investigators often find a large number of potential sources contributing low levels of many contaminants to the air.

The following information is presented for each IAQ phased evaluation:

- The basic goals, methodology, and sequence of the investigation, the information to be obtained, and the process of hypothesis development and testing, including criteria for decision-making about further data-gathering
- Any elements of the work that will require a time commitment from staff or residents, including information to be collected by the facility
- The schedule, cost, and work products, such as a written report, specifications, and plans for mitigation work, supervision of mitigation work, and training programs for staff
- Additional tasks (and costs) that may be part of solving the IAQ problem, but outside the scope of the contract (examples include medical examination of complainants, laboratory fees, and contractor's fees for mitigation work)

9.1.1.2 Source Assessment and Remediation

This IAQ phase includes either remote or on-site evaluation of known sources of airborne contaminants. When required, the IAQ team may also oversee necessary corrective measures. Examples of such projects include modification of smoking areas, decontamination of microbial growth surfaces, and response to pesticide spills.

The most efficient way to protect building air quality is through programs that prevent problems or address them in their earliest stages. In this regard IAQ aspects of facility design and property audits may be most effective if initiated during design and initial occupancy. Air quality precautions during facility design may include such steps as selecting and conditioning of materials to limit off-gassing, evaluation of proposed ventilation schemes, specifying remote sensing equipment, and preoccupancy testing of the building.

9.1.1.3 IAQ Risk Assessment

An investigation strategy based on evaluating building performance is used to solve a problem without necessarily identifying a particular chemical compound as the cause. The idea of testing the air to learn whether it is "safe" or "unsafe" is very appealing. However, most existing standards for airborne pollutants were developed for industrial settings where the majority of occupants are usually healthy adult men.

9.2 INDUSTRIAL/REMEDIATION PROCESS MONITORING

Through work efforts at industrial/remediation process sites, the basic precepts of industrial hygiene monitoring should be implemented from initial design through ongoing maintenance. The following is an example of an industrial hygiene scope of work to provide answers in process safety management.

9.2.1 Process Safety Management Example Scope of Work

Process safety management (PSM) is a new strategy of management controls where process limitations/hazards are identified and controlled so that injuries, spills, or hazards are minimized. It is mandated in 29 CFR 1910.119 and applies to manufacturing plants and other industries that are involved with large quantity chemical usage. The principal components required for PSM compliance will benefit any industry trying to manage chemical use. PSM is a complex system of controls, and in order to be successful in the implementation, upper management support and involvement is essential.

Standard operating procedures (SOPs) and O&M procedures must be reviewed. This review will focus on processes that involve chemicals that are potential air contaminants or other environmental media contaminants to the extent that health risk management is required. The PSM standard mandates that the SOPs have provisions for every operating phase, operating limitations, and safety and health considerations.

In general, the PSM requirements for operating procedures include the following:

- Operating procedures that must address steps for each operating phase, operating limits between various operations, safety and health considerations

- Copy of procedures that are readily accessible to employees

- Implementation of safe work practices

A review of spill containment capabilities must also occur to identify current spill containment areas and any area limitations. Coordination of spill responses with regulatory mandates and facility siting requirements should be addressed.

All hazard communication documentation must be reviewed. Hazard communication is a set of procedures to communicate the hazards potentially associated with storage, handling, shipping, or use of hazardous material. Communication is typically accomplished through material safety data sheets (MSDS), shipping papers, emergency response procedures, labels on containers, and training for employees. There must be open communication between all parties involved with hazardous material, including employer-employee, employee-employer, shipper-receiver, and the business-community. Specific procedures must be in place to facilitate this type of communication.

Past incidents will then be reviewed to determine any pattern or need for immediate change. By looking at past events, safeguards can be identified to defray accidents and releases in the future.

9.2.2 Provide List for Hazard and Operability Study

Current documents needed must be listed and acquired in order to complete a hazard and operability study (HAZOP). The current piping and instrumentation diagrams (P&ID) and other available design documents are a minimum requirement for review prior to initiation of the HAZOP.

Superfund Amendments and Reauthorization Act (SARA) Title III requires companies that use a large amount of chemicals to provide information on their plants and to participate in community emergency response plans. The SARA Title III reporting will be reviewed to identify any chemical uses that need to be addressed during the HAZOP study.

A process hazard analysis (PHA) provides employees and employers information to assist them in making good decisions about safety. The PHA focuses on the causes or contributing causes of potential accidents in the plant (fires, explosions, spills, releases, etc.). The PHA is directed toward the equipment, human actions, control panels, etc. that may cause failures in a system.

9.2.3 Process Hazard Analysis

Many variables must be considered when choosing the appropriate PHA technique to use, including the size of the plant, the number of processes, the types of processes, and the types of chemicals used. Many hazard analysis techniques can be used to determine the appropriate process safety aspects and physical distribution of a plant or facility. The following techniques may be used in performing a PHA.

Hazard Analysis Techniques
• Checklist
• "What if"
• HAZOP
• Failure mode and effect analysis
• Fault tree analysis
• Event tree analysis
• Distribution risk curves

9.2.3.1 Hazard and Operability Study

The HAZOP addresses each system and each element of a system that could deviate from normal operations and, thus, cause a hazard. A full assessment of each process is produced by looking at the hazards, consequences, causes, and personnel protection needed.

9.2.3.2 Failure Modes and Effects Analysis (FMEA)

FMEA involves the systematic rating of each process and the hazards involved. It is most effective when analyzing an individual process. It starts with a diagram of the design of the processes and identifies all the possible components that could fail (i.e., valves, hoses, pumps).

Components are analyzed for the following:

- Potential mode of failure
- Consequences of the failure
- Hazard class (high, moderate, low)
- Probability of failure
- Detection methods
- Compensating provisions/remarks

9.2.3.3 Fault Tree Analysis

Fault tree analysis is a graphical representation of all possible events of a hazard or hazards that may occur if a fault/breach in a process occurs. This method centralizes on the event that occurs, not so much the process. The graphical representation looks like a tree, and a probability is applied by the reviewer to each fault on the tree.

9.2.4 Design Analysis

Increasingly, designs must take into consideration safety and health requirements. These requirements include such topics as general safety, asbestos, lead, PCBs, volatile chemical/painting/other airborne chemical hazards, and hazardous waste generation. The following design documents should be prepared to provide the safety and health design team information within engineering and construction documents.

9.2.4.1 Site Safety and Health Plans

For sites, including future building sites, requiring a phase I and/or phase II environmental site assessment (ESA), a site safety and health plan must be developed for the investigating team if hazardous wastes are still significantly present on-site. This plan will

- Describe the minimum safety, health, and emergency response
- List anticipated major site tasks and operations
- Determine initial minimum levels of PPE

9.2.4.2 Health and Safety Design Analysis (HSDA)

During the design phase an HSDA is often developed. Site and/or hazard specific safety and health considerations and protective measures are outlined in the HSDA. The HSDA fully describes and justifies (using decision logic) the safety and health requirements to be included in the contract specifications or work plans.

In the event that a site safety and health plan has been approved for prior on-site investigative work, and the same contractor will be conducting the construction work, the HSDA only details the remaining construction work. If the remaining construction work does not pose additional hazards in excess of those discussed in the prior site safety and health plan, an HSDA will not be required.

9.2.4.3 Drawings

Drawings illustrate the following on an as-needed basis (i.e., either the drawings will contain this information or will illustrate how the information will be presented during a later design phase):

- The basic site layout, including the location of all existing utilities (e.g., water, sewer, gas, electrical)
- All potential interferences with utility routings, including any existing infrastructure that may pose an obstacle
- Site work zones for hazardous waste site work, including the EZ, the CRZ, and the SZ (When a work activity will be duplicated at varying locations on-site, an example of the placement of each zone is provided.)
- Schematic of decontamination facilities for both personnel and heavy equipment (Ideally these facilities should be in separate locations.)
- Location of all emergency facilities: eyewash stations, drench showers, first aid stations
- Routes to medical facilities
- Emergency and contingency plan staging areas
- Air-monitoring stations for background, area, and perimeter monitoring (If air monitoring is not required for any of these areas, the drawing notes must contain a negative declaration to this effect.)
- Waste collection, transfer, and disposal locations

9.2.4.4 Specifications

The contract requirements are fully developed from the HSDA and are described in a specification section. All minimum site-specific, task-specific, and hazard-specific procedures, precautions, and equipment determined necessary and described in the HSDA are clearly biddable and enforceable contractual requirements within the specification sections.

For hazardous waste sites, the site description and contamination characterization portion of the HSDA may be incorporated for information as a specification appendix.

In the event that a site safety and health plan has been approved for prior on-site investigative work, and the same contractor will be conducting the construction work:

- If the remaining remedial action construction work does not pose additional hazards in excess of those discussed in the prior approved plan, a safety and health specification section will not be required.
- If the remaining remedial action construction work does pose additional hazards in excess of those discussed in the prior approved plan, but the additional hazards can be adequately discussed in an addendum to the prior approved plan, a safety and health specification section will not be required.
- In lieu of requiring that the contractor write a safety and health specification section, a safety and health work plan may be required.
- If the remaining construction work does pose additional hazards in excess of those discussed in the prior approved site safety and health plan, and the additional hazards cannot be adequately discussed in an addendum to the prior approved site safety and health plan, a safety and health specification section will be required.

During any design analysis the chemical use and potential contaminant issues must be analyzed.

9.2.4.5 Design Analysis Example—Wastewater Treatment

The following is an example of the type of design analysis required to determine which chemicals could become air contaminants during process operations. This example is an excerpt from a design analysis that illustrates how process stream chemical activities are analyzed.

Wastewater Treatment

- Each reaction tank has a propeller mixer.
- Detention time in each tank is from 15 to 45 min.
- Reaction times for different wastes, reducing agents, temperatures, pH, and chromium concentrations.

After the Hexavalent Chromium Has Been Reduced to Trivalent Chromium

- The waste stream can be combined with other wastestreams containing common metals. These combined waste streams can be treated by the hydroxide precipitation process to remove chromium and the other common metals.
- Optimum levels for your metal waste treatment facilities may be slightly different due to the "mix" of wastes being treated.

Cyanide Destruction by Oxidation

The cyanide ion (CN^-) is extremely toxic and must be removed from metal wastes before discharge to sewers or the environment. Cyanide toxicity (poisoning) in humans is caused by an irreversible reaction with the iron in hemoglobin that results in loss of the blood's ability to transport oxygen.

If equipment failure occurs during the treatment of wastes containing cyanide, extremely toxic gases could be released. Therefore all tanks or pits used for cyanide destruction must be properly located and ventilated so that any gases produced will never enter an area occupied by people.

Cyanide Sources and Treatment

Cyanide compounds are used

- In plating solutions for copper, zinc, cadmium, silver, and gold
- In the immersion stripping of various electrodeposits
- In some activating solutions
- Complexed with heavy metals such as zinc, cadmium, silver, gold, copper, nickel, and iron

Flowing rinse waters after these production operations can become contaminated with cyanide. Cyanide may be present as the simple alkali cyanides of sodium or potassium.

For waste treatment control the two major groupings of cyanide compounds are segregated into

- Those that can be oxidized by chlorination (also known as cyanides amenable to chlorination)
- Those that resist oxidation by chlorination (also known as refracto-cyanides)

Both types are to toxic and must be removed or have their concentrations reduced below toxic levels prior to discharge of the rinse waters.

Those compounds that are readily oxidized by chlorination are cyanides of

- Sodium
- Potassium
- Cadmium
- Zinc

The copper cyanide complex is also considered amenable to chlorination, although longer reaction times are required (also for silver and gold cyanide complexes). The nickel cyanide complex is more resistant to chlorination than the copper complex, yet it can be considered as amenable to chlorination under extreme oxidizing conditions. Iron complexes (most commonly the sodium ferrocyanide salt) are not readily oxidizable by chlorine.

Pretreatment permits prescribe maximum limits for cyanides oxidizable by chlorination and for total cyanide (the sum of those that can be chlorinated and those that cannot).

- The average limits for amenable cyanides are 0.05 mg/l and for total cyanide 0.28 mg/l.
- The maximum concentrations for amenable and total cyanide are 0.65 mg/l and 1.2 mg/l.

Chemistry Involved

Waste treatment systems are designed to oxidize cyanide compounds with sodium hypochlorite (NaClO). The hypochlorite is purchased at a 14 to 15% by weight concentration. This solution has been produced by the reaction between chlorine and sodium hydroxide according to the reaction:

$$Cl_2 + 2NaOH \rightarrow NaClO + NaCl + H_2O$$

Sodium cyanate (NaCNO) is formed by the reaction of hypochlorite and cyanide.

- This reaction is actually a simplification of the chlorination of cyanide; very rapid intermediate reactions take place.
- Chlorine is freed from the hypochlorite by disassociation (a reversal of the formation of hypochlorite shown above).

The chlorine actually reacts with the cyanide to form cyanogen chloride according to

$$NaCN + Cl_2 \rightarrow CNCl + NaCl$$

The cyanogen chloride then reacts with the sodium hydroxide to form the cyanate according to

$$CNCl + 2NaOH \rightarrow NaCNO + NaCl + H_2O$$

The cyanogen chloride is a gas that has limited solubility at neutral or low pH values. At pH values of 10 or higher the reaction producing the cyanate is quite rapid for cyanides that are not too tightly complexed with heavy metals. At lower pH values not only is the reaction too slow, but also there is a good probability of liberating the cyanogen chloride gas to the atmosphere. This gas is

- A lacrymator (produces tears)
- Readily apparent to those nearby

Oxidation of cyanate is possible by chlorination where the reaction is

$$2NaCNO + 3NaOCl + H_2O \rightarrow 2CO_2 + N_2 + 3NaCl + 2NaOH$$

The carbon dioxide formed will react with alkali to produce carbonate. Any excess is liberated into the atmosphere.

- The nitrogen has a limited solubility and also escapes into the air.
- This second stage oxidation reaction is faster at pH values slightly lower than the initial reaction of cyanide to cyanate.

If the pH drops too low (below 7.0), then another reaction takes place with cyanate. This reaction will hydrolyze with water to produce ammonia compounds.

From this discussion pH as a critical factor must be defined. The rate of reaction, as well as the end products, depend upon maintaining proper pH values. The escape of cyanogen chloride is inhibited by high pH values.

Another impact of the oxidizing potential of pH at 10.5 to 11.0 is the oxidizer millivolts potential that results. Nitrogen chloride as an oxidizing gas essentially stops the reaction, so higher potentials of 650 mV do not solve the problem.

Operators must realize the extreme hazards that can develop and produce toxic cyanide gas whenever cyanide treatment processes are accidentally carried on at an acidic pH.

- Where significant levels of cyanides exist, extremely hazardous conditions can develop very quickly.
- Also when treating cyanides be sure to determine the pH at the start of the treatment process. The adjustment of pH and the various set points for the treatment system are to some extent dependent upon the concentration of cyanide and cyanate, and the initial pH of the wastestreams.
- Adjusting the pH downward to 9.0 for a treatment process when the initial pH of a cyanide solution is significantly higher could result in the evolution of toxic cyanide gas.

Cyanides complexed with zinc and cadmium react with the same rate as the simple sodium and potassium cyanides. Because the various other metal complexes are more resistant to oxidation by chlorination, long contact times are required.

In order to maintain a fast reaction rate, higher oxidizing potentials (additional excess chlorine) are employed. This method is most important for the complexes of silver and nickel.

When a cadmium cyanide complex is oxidized, the cadmium precipitates from solution as cadmium oxide. When oxidized, these metals precipitate from solution as hydroxides of copper, nickel, and zinc.

However, the presence of silver cyanide presents an additional concern. The silver cyanide complex when oxidized can precipitate as

- Silver chloride (a white precipitate)
- Silver oxide (a black precipitate)

Silver oxide is preferred; it is denser than silver chloride with the result that its removal from the wastewater is easier. Higher oxidizing conditions favor the oxide formation; lower oxidizing conditions favor the chloride formation.

The control devices are set to control the pH within the range of 10.5 to 11.5 for the first-stage reaction (cyanide to cyanate) and the second-stage reaction (cyanate to carbon dioxide and nitrogen). The control devices are set to maintain minimum oxidizing potentials of

- +600 mV in the first stage and
- +850 mV in the second stage

For the first stage reaction a minimum pH must be maintained. The top pH limit (11.5) is established.

Excessive alkali (as measured by pH) results in suppression of oxidation-reduction potential (ORP) values. The ORP value selected considers the need to have excess hypochlorite present to dominate the reaction. Sodium hypochlorite always contains free sodium hydroxide.

When the ORP controller requests the addition of hypochlorite, the pH can rise because of this simultaneous addition of alkali. When the pH rises, the ORP value decreases, establishing an out-of-control condition. Adding more hypochlorite results in the instrumentation requesting still more hypochlorite even though there is no cyanide present. This phenomenon is eliminated by the addition of acid whenever the pH rises too high.

CHAPTER 10

Occupational Health—
Air-Monitoring Strategies

This chapter discusses the current OSHA and NIOSH protocols and their applicability. It includes in tabular form the air-monitoring strategies that may be approved in the future.

Once substantial employee exposure is a potential concern, such exposure needs to be measured. The next step is to select a maximum-risk employee. When there are different processes where employees may be exposed to the chemical of concern, a maximum-risk employee should be selected for each work operation. The action level should usually be the OSHA PEL.

Selection of the maximum-risk employee requires professional judgment. The best procedure for selecting the maximum risk employee is to observe employees and select the person closest to the chemical source. Employee mobility may affect this selection, e.g., if the closest employee task requires mobility, this worker may not be the maximum risk employee. Air movement patterns and differences in work habits will also affect selection of the maximum risk employee.

When many employees perform essentially the same task, a maximum-risk employee cannot be selected. In this circumstance it is necessary to resort to random sampling of the group of workers. The objective is to select a subgroup of adequate size so that there is a high probability that the random sample will contain at least one worker with high exposure. The number of persons in the group influences the number that needs to be sampled to ensure that at least one individual from the highest 10% exposure group is contained in the sample. For example, to have 90% confidence in the results, if the group size is 10, 9 will be sampled; for 50, only 18 need to be sampled.

If measurement shows exposure to the chemical at or above the action level or the STEL, all other employees who may be exposed at or above the action level or STEL need to be identified. Measurement or otherwise accurate characterization of their exposure must be accomplished.

Whether representative monitoring or random sampling is conducted, the purpose remains the same: to determine if the exposure of any employee is above the action level. If the exposure of the most exposed employee is less than the action level and the STEL, regardless of how the employee is identified, then it is reasonable to assume that measurements of exposure of the other employees in that operation would be below the action level and the STEL.

10.1 EXPOSURE MEASUREMENTS

There is no "best" measurement strategy for all situations. Some elements to consider in developing a strategy are as follows:

- Availability and cost of sampling equipment
- Availability and cost of analytic facilities
- Availability and cost of personnel to take samples
- Location of employees and work operations
- Intraday and interday variations in the process
- Precision and accuracy of sampling and analytic methods
- Number of samples needed

Personal air-sampling pumps should be both sturdy and difficult to tamper with during use. Variations in flow rate as tubing to and from the pump is moved must be adjusted through electronic compensation that maintains a calibrated flow rate (Figure 10.1).

The authors prefer personal sampling pumps that do *not* have attached rotameters; these rotameters cannot be used to calibrate the pumps or to guarantee flow rate. Rather, these attached rotameters are a rough measure of flow rates that unfortunately are often thought of as calibration devices.

10.2 STEL SAMPLING

Samples taken for determining compliance with the STEL differ from those that measure the TWA concentration in important ways. STEL samples are best taken nonrandomly using all available knowledge relating to the area, the individual, and the process to obtain samples during periods of maximum expected concentrations.

At least three measurements on a shift are generally needed to spot gross errors or mistakes; however, only the highest value represents the STEL. If an operation remains constant throughout the workshift, a much greater number of samples would need to be taken over the 32 discrete nonoverlapping periods in an 8-h workshift to verify compliance with a STEL.

Figure 10.1 Personal air sampling pump. This model has an internal secondary standard that calibrates the pump continuously and is checked against a primary standard only once a month. (MSA—Escort ELF)

10.3 EXPOSURE FLUCTUATIONS

Interday and intraday fluctuations in employee exposure are mostly influenced by the physical processes that generate formaldehyde and the work habits of the employee. Hence, in-plant process variations influence the determination of whether additional controls need to be imposed. Measurements in which employee exposure is low on a day that is not representative of worst conditions may not provide sufficient information to determine whether additional engineering controls will be installed to achieve the PELs.

The person responsible for sampling must be aware of systematic changes that will negate the validity of the sampling results. Systematic changes in exposure concentration for an employee can occur due to

- The employee changing patterns of movement in the workplace
- Closing plant doors and windows
- Changes in ventilation from season to season
- Decreases in ventilation efficiency or abrupt failure of engineering control equipment
- Changes in the production process or work habits of the employee

Any of these changes, if they may result in additional exposure that reaches the next level of action, requires the performance of additional monitoring to reassess employee exposure.

10.4 AIR-SAMPLING PUMP USER OPERATION

10.4.1 Pump Donning

- Attach the collection device (filter cassette, charcoal tube, etc.) to the shirt collar or as close as practical to the nose and mouth of the employee, i.e., in a hemisphere forward of the shoulders with a radius of approximately 6 to 9 in.
- The inlet should always be in a downward vertical position to avoid gross contamination. Position the excess tubing so that it does not interfere with normal activity.
- Turn on the pump and record the starting time.
- Observe the pump operation for a short time after starting to make sure the pump is operating correctly. Record the required information.

10.4.2 Pump Checking

- Check pump every 2 h.
- More frequent checks may be necessary with heavy filter loading.
- Ensure that the sampler is still assembled properly and that the hose has not become pinched or detached from the cassette or the pump.
- For filters observe for symmetrical deposition, fingerprints, large particles, etc.
- Record the flow rate.
- Periodically monitor the employee throughout the workday to ensure that sample integrity is maintained and cyclical activities and work practices are identified.

10.4.3 Pump Doffing

- *Before* removing the pump at the end of the sample period, if there is a pump rotameter, check the flow rate to ensure that the rotameter ball is still at the calibrated mark. If the ball is no longer at the mark, record the pump rotameter reading. Keep in mind that the pump-mounted rotameters are not as accurate as separate precision rotameters.

- Turn off the pump and record the ending time. Measure and record the flow rate using the calibration device. Do not rely on the pump rotameter.
- Remove the collection device from the pump and seal the filter cassette or tube (Figures 10.2 and 10.3). The seal should be attached across the sample inlet and outlet to prevent tampering.
- Prepare the samples for mailing to the laboratory for analysis.
- Recalibrate pumps after each day of sampling (before charging). Calibrate personal sampling pumps (Figure 10.4) before and after each day of sampling, using a bubble meter method (electronic or mechanical) or the precision rotameter

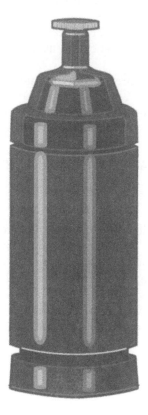

Figure 10.2   Capped 25 mm filter cassette. (SKC)

Figure 10.3   Capped 37 mm filter cassette. (SKC)

method (Figures 10.5 and 10.6) (using a precision rotameter that has been calibrated against a bubble meter).

For unusual sampling conditions, such as wide temperature and barometric pressure differences from calibration conditions, you will need to calculate the effect of these conditions on your ultimate sampling results.

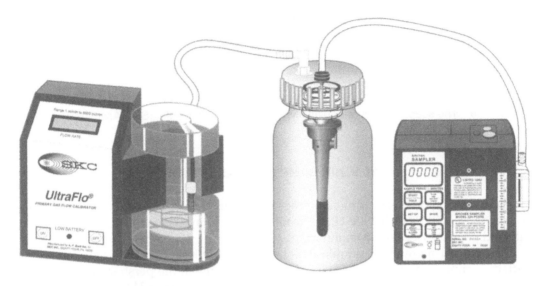

Figure 10.4 Use of calibration chamber. (SKC)

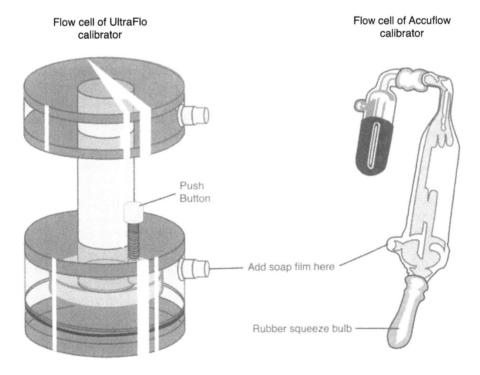

Flow cell of UltraFlo
calibrator

Flow cell of Accuflow
calibrator

Push
Button

Add soap film here

Rubber squeeze bulb

Figure 10.5 Calibrator flow cells. (SKC)

Figure 10.6   The soap film calibrator digitally records liters per minute. (SKC Accuflow®)

10.5 AIR SAMPLING—ASBESTOS

Sampling and analysis of the airborne concentration of asbestos fibers are performed in accordance with 29 CFR 1926.1101. The following is sample specification language to initiate and provide asbestos air monitoring.

Personal air-monitoring samples will be taken for at least 25% of the workers in each shift, or a minimum of two, whichever is greater. Results of the personal samples will be posted at the job site and made available to the work force within 24 h.

The asbestos site supervisor will maintain a fiber concentration inside regulated work areas that will be equal to or less than 0.1 f/cc expressed as an 8-h TWA during the conduct of the asbestos abatement. If fiber concentration rises above 0.1 f/cc, work procedures will be investigated to determine the cause.

Fiber concentration may exceed 0.1 f/cc, but will not exceed 1.0 f/cc expressed as an 8-h TWA. Workers must not be exposed to an airborne fiber concentration in excess of 1.0 f/cc, averaged over a sampling period of 30 min.

Should either an environmental concentration of 1.0 f/cc expressed as an 8-h TWA or a personal excursion concentration of 1.0 f/cc expressed as a 30-min sample occur inside the regulated work area, the asbestos site supervisor will

- Stop work immediately
- Notify the responsible party (owners or their agents)
- Implement additional engineering controls and work practice controls to reduce airborne fiber levels below prescribed limits in the work area.
- Not restart work until authorized by the responsible party

Air monitoring results at the 95% confidence level will be calculated. The asbestos contractor will provide an independent testing laboratory with qualified analysts and appropriate equipment to conduct sample air analyses using the methods prescribed in 29 CFR 1926.58 and NIOSH Pub No. 84-100 and Method 7400. Sampling is performed in accordance with 29 CFR 1926.1101.

For personal sampling required by 29 CFR 1926.1101, NIOSH Pub No. 84-100 and Method 7400 will be used for sampling and PCM analysis. For environmental quality control and final air clearance NIOSH Pub No. 84-100 and Method 7400 (PCM) with optional confirmation of results by NIOSH Pub No. 84-100 and Method 7402 (TEM) [the mandatory EPA TEM Method specified at 40 CFR Part 763] will be used.

For environmental and final clearance samples, sampling will be conducted at a sufficient velocity and time to collect a sample volume necessary to establish the limit of detection of the method used at 0.005 f/cc. Asbestos fiber concentration confirmation of the total fiber concentration results of environmental, quality assurance, and final air clearance samples, collected and analyzed by NIOSH Pub No. 84-100 and Method 7400, may be conducted.

10.5.1 Sampling Prior to Asbestos Work

Baseline air sampling will be established immediately prior to the initiation of work in each abatement site. The background will be established by performing sampling in similar, but uncontaminated sites in the building. Preabatement air samples will be collected at a minimum of three locations. These locations will be outside the building; inside the building, but outside the abatement perimeter; and inside each abatement perimeter. One sample will be collected for each of the three laboratories' floor space (Figure 10.7).

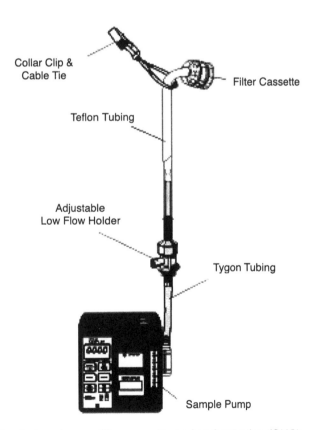

Figure 10.7 *Sampling train using pre-filter cassette and sorbent tube. (SKC)*

10.5.2 Sampling during Asbestos Abatement Work

The asbestos contractor and asbestos site supervisor will provide and/or oversee personal air sampling as indicated in 29 CFR 1926.1101, State of New York and local requirements, and in accordance with this plan.

Sampling will be conducted at least once each day in the laboratories where work is occurring, close to the work in the asbestos-regulated space. If airborne fiber levels have exceeded background or 0.01 f/cc, whichever is greater, all work will be stopped immediately, and the responsible party notified. The condition causing the increase will be corrected. Work will not restart until authorized by the responsible party.

Where the construction of a containment is not required, after initial TWA airborne fiber concentrations are established, and provided the same type of work is being performed, PCM sampling will be conducted at the boundary of the asbestos-regulated work area in such locations and at such frequency as recommended by the on-site air-monitoring professional.

10.5.3 Sampling after Final Cleanup (Clearance Sampling)

Prior to conducting final air clearance monitoring, a final visual inspection of the asbestos contractor's final cleanup of the abated asbestos-regulated work will occur. Final clearance air monitoring will not begin until acceptance of this final cleaning by the responsible parties. Ongoing monitoring may also be used to prove that final clearance levels were never exceeded.

Final clearance sampling of airborne fibers will be accomplished using aggressive air sampling techniques as defined in EPA 560/5-85-024 or as otherwise required by federal or state requirements. The sampling and analytical method used will be mandatory [NIOSH Pub No. 84-100 and Method 7400 (PCM)] with optional confirmation of results by NIOSH Pub No. 84-100 and Method 7402 (TEM). [The mandatory EPA TEM method is specified at 40 CFR Part 763.]

The final clearance air samples will be collected and analyzed as indicated by NIOSH Pub No. 84-100 and Method 7400 (PCM) or the EPA mandatory method 40 CFR Part 763 (TEM).

10.5.4 NIOSH Method

For PCM sampling and analysis using NIOSH Pub No. 84-100 and Method 7400, the fiber concentration inside the abated asbestos-regulated work area for each airborne sample is less than 0.01 f/cc. Decontamination of the abated asbestos-regulated work area will be considered complete when every PCM final clearance sample is below the clearance limit.

- If any sample result is greater than 0.01 f/cc, the asbestos fiber concentration from that same filter will be reconfirmed using NIOSH Pub No. 84-100 and Method 7402 (TEM).
- If any confirmation sample result is greater than 0.01 f/cc, then abatement is incomplete, and recleaning is required.

Upon completion of any required recleaning, resampling with results to meet the above clearance criteria is required.

10.5.5 Air Sampling Documentation

Air sample fiber counting will be completed and results provided within 24 h after completion of a sampling period. The responsible party will be notified immediately of any airborne levels of asbestos fibers in excess of established requirements. Written sampling results will be provided within 5 working days of the date of collection. The written results will be signed by the testing laboratory analyst, the testing laboratory manager, the asbestos site supervisor, and the air-monitoring professional. The air-sampling results will be documented on the asbestos contractor's daily air monitoring log. The daily air monitoring log will contain the following information for each sample:

- Date sample collected
- Date sample analyzed
- Sample number
- Sample type (p = personal, a = area, c = abatement clearance, irwa = inside regulated work, orwa = outside regulated work, du = decontamination unit, lou = load-out unit, at = access tunnel)
- Sample period (start time, stop time, elapsed time [minutes])
- Sampling pump manufacturer—model and serial number
- Total air volume sampled (liters)
- Sample results (fibers per cubic centimeter and structures per square millimeter)
- Location/activity/name where sample collected
- The calibration method used to calibrate the sampling pumps
- The name and location of the laboratory conducting the sample analyses
- Printed name, signature, and date block for the industrial hygienist who conducted the sampling
- The review verifying the accuracy of the information

Table 10.1   Formula for Calculation of the 95% Confidence Level (Reference: NIOSH 7400)

$$f/cc~(01.95\%~CL) = X + [(X) \times (1.645) \times (CV)]$$

where

$$X = [(E)~(AC)]/[(V)~(1000)]$$

$$E = [(F/Nf) - (B/Nb)]/Af$$

- CV = the precision value; 0.45 will be used unless the analytical laboratory provides the responsible party's representative with documentation (round robin program participation and results) that the laboratory's precision is better
- AC = effective collection area of the filter in square millimeters
- V = air volume sampled in liters
- E = fiber density on the filter in fibers per square millimeter
- F/Nf = total fiber count per graticule field
- B/Nb = mean field blank count per graticule field
- Af = graticule field area in square millimeters

Table 10.2   NIOSH Method 7400: PCM Environmental Air Sampling Protocol (Nonpersonal)

Sample Location	Minimum No. of Samples	Filter Pore Size[1] (μm)	Min. Vol.[2] (l)	Sampling Rate (l/min)
Inside Abatement Area[3,4]	5/1500 ft^2	0.45 or 0.8	1500	2–10
Each Room in Abatement Area Less than 1500 ft^2	1	0.45 or 0.8	1500	2–10
Field Blank	2	0.45 or 0.8	0	0
Laboratory Blank	1	0.45 or 0.8	0	0

Notes:

1. Type of filter is mixed cellulose ester.
2. Ensure detection limit for PCM analysis is established at 0.005 f/cc.
3. One sample should be added for each additional 1500 ft^2.
4. No less than five samples are to be taken per abatement area, plus two field blanks.

Table 10.3   EPA AHERA Method: TEM Air Sampling Protocol

Sample Location	Minimum No. of Samples	Filter Pore Size[1] (μm)	Min. Vol.[2] (l)	Sampling Rate (l/min)
Inside Abatement Area	5	0.45	1199	2–10
Outside Abatement Area	5	0.45	1199	2–10
Field Blank	2	0.45	0	0
Laboratory Blank	1	0.45	0	0

Notes:

1. Type of filter is 0.8 μm mixed cellulose ester.
2. The detection limit for TEM analysis is 70 s/mm^2.

10.5.6 Asbestos Exposure Monitoring (29 CFR 1910.1001 and 29 CFR 1926.1101)

Asbestos exposure monitoring is a component of abatement work and may be required at any time asbestos exposure is a concern. Employee exposure means that exposure to airborne asbestos that would occur if the employee were *not* using respiratory protective equipment.

Employee exposures are determined using breathing zone air samples (8-h TWA PELs and 30-min STELs). Representative 8-h TWA employee exposures are one or more samples representing full-shift exposures for

- Each shift
- Each employee
- Each job classification
- In each work area

Representative 30-min STELs are one or more samples representing 30-min exposures (operations that are most likely to produce exposures above the excursion limit) for

- Each shift
- Each job classification
- In each work area

10.5.7 Initial Monitoring

Begin initial monitoring immediately before/at the initiation of the operation (1926.1101):

- When employees *may reasonably be expected to be* exposed to airborne concentrations at or above the action level and/or excursion limit (1910.1001) made from breathing zone air samples that are representative of the 8-h TWA and 30-min STELs of each employee.
- Representative 8-h TWA employee exposures are determined on the basis of one or more samples representing full-shift exposure for employees in each work area.
- Representative 30-min STELs are determined on the basis of one or more samples representing 30-min exposures associated with operations that are most likely to produce exposures above the excursion limit for employees in each work area.

10.5.8 Historical Documentation for Initial Monitoring

Historical documentation for initial monitoring may be used when:

- Monitoring satisfies all the above requirements and the analytical requirements listed in 29 CFR 1910.1001 and 1926.1101
- Documentation has occurred within the past calendar year (12 months)

10.5.9 Objective Data for Initial Monitoring

Objective data for initial monitoring may be used when asbestos is not capable of being released in airborne concentrations at or above the action level and/or excursion limit under the expected conditions of processing, use, or handling. No initial monitoring is then required.

10.6 CRYSTALLINE SILICA SAMPLES ANALYZED BY X-RAY DIFFRACTION (XRD)

10.6.1 Air Samples

Respirable dust samples for quartz, cristobalite, and tridymite are analyzed by XRD. XRD is the preferred analytical method due to its sensitivity, its minimum requirements for sample preparation, and its ability to identify polymorphs (different crystalline forms) of free silica. For the analysis of crystalline free silica by XRD, the particle size distribution of the samples must be matched as closely as possible to the standards, which is best accomplished by collecting a respirable sample. Respirable dust samples are collected on a tared low ash PVC filter using a 10-mm nylon cyclone at a flow rate of 1.5–1.9 l/min.

- A sample not collected in this manner is considered a total dust (or nonrespirable) sample. Industrial hygienists are discouraged from submitting total dust samples because an accurate analysis cannot be provided by XRD for such samples.
- If the sample collected is nonrespirable, the laboratory must be advised on the chain-of-custody record.

Quartz (also cristobalite and tridymite) is initially identified by its major (primary) XRD peak. A few substances also have peaks near the same location, and it is necessary to confirm quartz (also cristobalite or tridymite) using secondary and/or tertiary peaks. To assist the analyst in identifying interference, the industrial hygienist should provide information concerning potential presence of other substances in the workplace. The following substances should be noted:

- Aluminum phosphate
- Feldspars (microcline, orthoclase, plagioclase)
- Graphite
- Iron carbide
- Lead sulfate
- Micas (biotite, muscovite)
- Montmorillonite
- Potash
- Sillimanite
- Silver chloride
- Talc
- Zircon (zirconium silicate)

A sample weight and total air volume should accompany all filter samples. Sample weights of 0.5–3.0 mg are preferred. Do not submit samples unless their weight, or the combined weights of all particulate filters representing an individual exposure, exceed 0.04 mg.

If heavy sample loading is noted during the sampling period, change the filter cassette to avoid collecting a sample with a weight greater than 5.0 mg.

10.6.1.1 Laboratory Results for Air Samples

- Percent quartz (or cristobalite): Applicable for a respirable sample in which the amount of quartz (or cristobalite) in the sample was confirmed.
- Less than or equal to value in units of percent: Less than or equal to values are used when the adjusted 8-h exposure is found to be less than the PEL, based on the sample's primary diffraction peak.
- The value reported represents the maximum amount of quartz (or cristobalite) that could be present.
- The presence of quartz (or cristobalite) was not confirmed using secondary and/or tertiary peaks in the sample because the sample could not be in violation of the PEL.

- Approximate values in units of percent: The particle size distribution in a total dust sample is unknown and error in the XRD analysis may be greater than for respirable samples. Therefore, for total dust samples, an approximate result is given.
- Nondetected:
  - —A sample reported as nondetected indicates that the quantity of quartz (or cristobalite) present in the sample is not greater than the detection limit of the instrument. The detection limit is usually 10 mg for quartz and 30 mg for cristobalite.
  - —If less than a full-shift sample was collected, the industrial hygienist should evaluate a nondetected result to determine whether adequate sampling was performed.
  - —If the presence of quartz (or cristobalite) is suspected in this case, the industrial hygienist may want to sample for a longer period of time to increase the sample weights.

10.6.2 Bulk Samples

Bulk samples should be submitted for all silica analyses, if possible. Bulk samples have the following purposes:

- To confirm the presence of quartz or cristobalite in respirable samples or to assess the presence of other substances that may interfere in the analysis of respirable samples
- To determine the approximate percentage of quartz (or cristobalite) in the bulk sample
- To support hazard communication inspections

A bulk sample must be representative of the airborne free silica content of the work environment sampled; otherwise, it has no value.

The laboratory's order of preference for bulk samples for an evaluation of personal exposure is as follows:

- A high-volume respirable area sample
- A high-volume area sample
- A representative settled-dust (rafter) sample (This is the most practical option. In certain operations it may be very difficult to collect enough material using high-volume sampling to be used as a bulk sample.)
- A bulk sample of the raw material used in the manufacturing process (most practical if used for hazard communication inspections)

The type of bulk sample submitted to the laboratory should be stated on the OSHA-91 form and cross-referenced to the appropriate air samples.

A reported bulk sample analysis for quartz (also cristobalite or tridymite) will be semiquantitative because

- Errors are associated with bulk sampling.
- The XRD analysis procedure requires a thin layer deposition for an accurate analysis.
- The error for bulk samples analyzed by XRD is unknown because the particle size of nonrespirable bulk samples varies from sample to sample.

10.6.3 Sample Calculations for Crystalline Silica Exposures

Where the employee is exposed to combinations of silica dust (i.e., quartz, cristobalite, and tridymite), the additive effects of the mixture will be considered. For the PEL calculation specified in 29 CFR 1910.1000, Table Z-3, the percent silica will be determined by doubling the percentage of cristobalite and/or tridymite and adding it to the percentage of quartz, according to the following formula. The PEL mixture pertains to the respirable fraction.

$$10 \text{ mg/m}^3 \text{ PEL} = \% \text{ quartz} + 2(\% \text{ cristobalite}) + 2(\% \text{ tridymite}) + 2$$

10.6.4 Sample Calculation for Silica Exposure

Two consecutive samples from the same employee taken for a combined exposure to silica dusts have the following results:

Sample	Total Period	Respirable Volume (l)	Respirable Weight (mg)	Laboratory Concentration (mg/m³)	Percentage
A	238	405	0.855	2.1	5.2 quartz 2.3 cristobalite ND tridymite
B	192	326	0.619	1.9	4.8 quartz 1.7 cristobalite ND tridymite
Total	430	731	1.474		

Step 1. Calculate the percentage of quartz, cristobalite, and tridymite in the respirable particulate collected.

Quartz: $5.2(0.855/1.474) + 4.8(0.619/1.474) = 3.0 + 2.0 = 5.0\%$

Cristobalite: $2.3(0.855/1.474) + 1.7(0.619/1.474) = 1.3 + 0.7 = 2.0\%$

Step 2. Calculate the PEL for the mixture.

$$\text{mg/m}^3 \text{ PEL} = \% \text{ quartz} + 2(\% \text{ cristobalite}) + 2(\% \text{ tridymite}) + 2$$
$$= 10/[5.0 + 2(2.0) + 2(0) + 2] = 10/11.0 = 0.91 \text{ mg/m}^3$$

Step 3. Calculate the employee's exposure.

$$\text{Exposure} = (\text{Sample wt. A} + \text{Sample wt. B})/\text{Total volume}$$
$$= (0.855 + 0.619)/0.731 = 2.0 \text{ mg/m}^3$$

Step 4. Adjust (where necessary) for less than 8-h TWA sampling period.

$$= (2.0 \text{ mg/m}^3)[(430 \text{ min})/(480 \text{ min})] = 1.8 \text{ mg/m}^3$$

Step 5. Calculate the severity of the exposure.

$$1.8 \text{ mg/m}^3/(0.91 \text{ mg/m}^3) = 2.0$$

10.7 METALS—WELDING

Where two or more of the following analytes are requested on the same filter, an ICP analysis may be conducted. However, the industrial hygienist should specify the metals of interest in the event samples cannot be analyzed by the Inductively Coupled Plasma (ICP) method. A computer printout of the following analytes may be reported:

- Antimony
- Beryllium
- Chromium
- Cobalt
- Copper
- Iron
- Lead
- Manganese
- Molybdenum
- Nickel
- Vanadium
- Zinc

If requested, the laboratory can analyze for "solder-type" elements, such as

- Antimony
- Beryllium
- Cadmium
- Copper
- Lead
- Silver
- Tin
- Zinc

Note: Samples taken during abrasive blasting operations are no longer analyzed by ICP because of difficulties with heavy loadings. These samples can be analyzed by atomic absorption spectrometry (AAS) for specific metals (i.e., Pb, Cd, Cr, Fe).

10.8 GENERAL TECHNIQUE FOR WIPE SAMPLING

10.8.1 Filter Media and Solvents

Consult a CIH for appropriate filter media and solvents. Dry wipes may be used. Solvents are not always necessary, but may enhance removal.

Direct skin wipes should not be taken when high skin absorption of a substance is expected. Under no conditions should any solvent other than distilled water be used on the skin, on personal protective gear that comes into direct contact with the skin, or on surfaces that come into contact with food or tobacco products.

Generally, two types of filters are recommended for taking wipe samples:

1. *Glass fiber filters (GFF) (37 mm)* are usually used for materials that are analyzed by high performance liquid chromatography (HPLC) and often for substances

analyzed by gas chromatography (GC). The OCIS chemical sampling information specifies when GFFs are to be used.

2. Paper filters are generally used for metals. For convenience the Whatman smear tab (or its equivalent) or polyvinyl chloride filters for substances that are unstable on paper-type filters are commonly used.

For some volatile organics, charcoal pads are used.

10.8.2 Wipe Sampling Procedures

Preloading a group of vials with appropriate filters is a convenient method. (The Whatman smear tabs should be inserted with the tab end out.) Always wear clean plastic gloves when handling filters. Gloves should be disposable and should not be powdered.

Follow these procedures for taking wipe samples:

- If multiple samples are to be taken at the worksite, prepare a rough sketch of the areas or rooms to be wipe sampled.
- Use a new set of clean, impervious gloves for each sample to avoid contamination of the filter by the hand (and the possibility of false positives) and to prevent contact with the substance.
- Withdraw the charcoal pad from the vial. If a damp wipe sample is desired, moisten the filter with distilled water or other solvent as recommended by the chosen analytical laboratory.
- **Caution:** Skin, PPE, or surfaces that come into contact with food or tobacco products must be wiped either *dry* or with distilled water, never with organic solvents. Skin wipes should not be done for materials with high skin absorption. It is recommended that hands and fingers are the only skin surfaces wiped. Before any skin wipe is taken, explain why you want the sample and ask the employee about possible skin allergies to the chemicals in the sampling filter or medium. If the employee refuses, do not force the issue.
- Wipe a section of the surface to be sampled using a template with an open area of exactly 100 cm^2.
    - —For surfaces smaller than 100 cm^2 use a template of the largest size possible. Be sure to document the size of the area wiped. For curved surfaces the wiped area should be estimated as accurately as possible and then documented.
    - —Maximum pressure should be applied when wiping.
    - —To ensure that all of the partitioned area is wiped, start at the outside edge and progress toward the center by wiping in concentric squares of decreasing size.
    - —If the filter/pad dries out during the wiping procedure, rewet the filter (for wetted filters only).
- Without allowing the filter/pad to come into contact with any other surface, fold the filter/pad with the exposed side in, then fold it over again.
- Place the filter/pad in a sample vial.
- Cap and number the vial.
- Note the number at the sample location on the sketch. Include notes with the sketch giving any further description of the sample.

At least one blank filter/pad treated in the same fashion, but without wiping, should be submitted for each sampled area.

10.8.3 Special Technique for Wipe Sampling with Acids and Bases

When examining surfaces for contamination with strong acids or bases (e.g., hydrochloric acid, sodium hydroxide), pH paper moistened with water may be used. However, results should be viewed with caution due to potential interference.

10.8.4 Direct-Reading Instruments

For some types of surface contamination direct-reading instruments may be used (e.g., mercury sniffer for mercury).

10.8.5 Aromatic Amines

Screening may determine the precise areas of carcinogenic aromatic amine contamination. This is an optional procedure.

10.8.6 Special Considerations

Due to their volatility, most organic solvents are not suitable for wipes. Other substances are not stable enough as samples to be wipe-sampled reliably. If necessary, judge surface contamination by other means, (e.g., by use of detector tubes, photoionization analyzers, or similar instruments).

Some substances should have solvent added to the vial as soon as the wipe sample is placed in the vial (e.g., benzidine). These substances are indicated with an "X" next to the solvent notation in the ACGIH biological exposure index (BEI).

Do not take surface wipe samples on skin if:

- OSHA or ACGIH shows a "skin" notation and the substance has
  —A skin LD_{50} of 200 mg/kg or less
  —An acute oral LD_{50} of 500 mg/kg or less
- The substance
  —Is an irritant,
  —Causes dermatitis or contact sensitization
  —Is termed corrosive

10.8.6.1 Fluorescent Screening for Carcinogenic Aromatic Amines

As in the case of routine wipe sampling, wear clean, disposable, impervious gloves. Wipe an area of exactly 100 cm^2 with a sheet of filter paper moistened in the center with 5 drops of methanol.

- After wiping the sample area, apply 3 drops of fluorescamine (a visualization reagent supplied by the laboratory upon request) to the contaminated area of the filter paper.
- Place a drop of the visualization reagent on an area of the filter paper that has not come into contact with the surface. This step marks a nonsample area or blank on the filter paper adjacent to the test area.
- After a reaction time of 6 min, irradiate the filter paper with 366-nm ultraviolet light.

- Compare the color development of the sample area with the nonsample or blank area. A positive reaction shows yellow discoloration that is darker than the yellow color of the fluorescamine blank.
- A discoloration indicates surface contamination—a possible aromatic amine carcinogen. Repeat a wipe sampling of the contaminated areas using the regular surface contamination procedure.

The following compounds are some of the suspected carcinogenic agents that can be detected by this screening procedure:

- 4,4'-Methylene bis(2-chloroaniline)
- Benzidine
- Alpha-napthylamine
- Beta-napthylamine
- 4-Aminobiphenyl

10.8.6.2 Alternate Screening Methods for Aromatic Amines

OSHA is testing commercially available kits with wipe pads that contain an aromatic amine indicator. Preliminary evaluations show them to be an adequate screening tool. Their detection limit is approximately 5.0 mg/100 cm^2. These kits are more convenient than the fluorescent procedure outlined above, and they eliminate the added hazard of handling fluorescamine.

The following compounds are among the suspected agents that can be detected through this screening procedure:

- Methylene dianiline (MDA)
- 4,4'-Methylene bis(2-chloroaniline)
- Benzidine
- Alpha-napthylamine
- Beta-napthylamine
- 4-Aminobiphenyl
- o-Toluidine
- Aniline
- 2,4-Toluenediamine
- 1,3-Phenylenediamine
- Napthylenediamine
- 2,4-Xylidine
- o-Chloroaniline
- 3,4-Dichloroaniline
- p-Nitroaniline

CHAPTER 11

Monitoring for Toxicological Risk

This chapter gives monitoring information when individuals are at greater risk from toxic effects. These individuals may be immune compromised due to either disease or age (i.e., newborns, children to age three, adults approaching old age). Special emphasis is given to monitoring that can be accomplished without emotional duress.

Exposure monitoring may be conducted for chemical/physical agents to determine toxicological and carcinogenic chemical risk or other exposure parameters. The need for monitoring is based on the adequacy of the historical exposure data and the nature of the stress. Exposure monitoring is conducted to characterize personnel exposure where there is little or no database information or when operation/process conditions have changed. The data are used to assess the need for engineering and/or administrative controls or the use of protective equipment.

Employees should be notified of the results as soon as possible after the data have been collected and evaluated. Industrial hygiene reports are issued after completion of the sample collection and analysis portions of the monitoring surveys. Monitoring is performed as specified in OSHA regulations 29 CFR 1910.1450 (laboratory), 29 CFR 1910.120 (hazardous waste operations and emergency response), 29 CFR 1926.58 (asbestos, tremolite, anthophylite, and actinolite), 29 CFR 1910.1028 (benzene), 29 CFR 1910.1027 (cadmium), and any other situations deemed appropriate.

11.1 TYPES OF SAMPLING

Four types of samples are taken to determine potential or actual exposures to chemical stresses within the workplace: long-term (8–12 h) TWA samples, short-term samples (5 to 60 min), area samples, and wipe tests.

11.1.1 Long-Term Samples

Long-term samples are collected to determine average exposures throughout the typical work shift. Usually four to six samples are sufficient to assess the exposure potential for a job classification.

11.1.2 Short-Term Samples

Short-term samples are collected to determine peak exposure potential during the work shift. When long-term samples are higher than expected, short-term samples can help identify the specific tasks that possibly produced the high long-term average. Short-term samples are collected during an operation that lasts from 5 min to 1 h to determine the average exposure potential for the task. Short-term samples are collected for tasks such as groundwater/soil sampling, underground storage tank inspections, analytical laboratory operations, and asbestos sampling/inspections/abatement observation activities. Controls can then be applied to the situation to reduce the exposure potential for the short-term task as well as the long-term exposure.

11.1.3 Area Samples

Area samples are collected to identify background levels of airborne contaminants. They are useful in identifying contaminants from vents, open tanks, and other fugitive emissions. From a practical standpoint they are not generally used to determine personal exposures for unusual circumstances.

For sites defined as hazardous waste sites, additional area monitoring may be required. Monitoring must be conducted before site entry at uncontrolled hazardous waste sites to identify conditions immediately dangerous to life and health, such as oxygen-deficient atmospheres and areas where toxic substance exposures are above permissible limits.

Accurate information on the identification and quantification of airborne contaminants is useful for

- Selecting PPE
- Delineating areas where protection and controls are needed
- Assessing the potential health effects of exposure
- Determining the need for specific medical monitoring

After a hazardous waste cleanup operation begins, periodic monitoring of those personnel who are likely to have higher exposures must be conducted to determine if they have been exposed to hazardous substances in excess of PELs. Monitoring must also be conducted for any potential condition that is immediately dangerous to life and health or for higher exposures that may occur as a result of new work operations.

11.1.4 Wipe Samples

Wipe samples are collected to evaluate the tracking of chemicals. Office areas and equipment must be free of potentially harmful chemicals.

11.2 QUALITY CONTROL

Routine quality control procedures will be an integral part of the sampling and analysis procedures. Use of blanks, spikes, and routine calibration of equipment will be included in the quality control of the data.

11.3 EXPOSURE EVALUATION CRITERIA

There are many sources for general exposure guidelines. The relevant evaluation criteria usually include the OSHA PELs, the recommended threshold limit values (TLVs) of the ACGIH, and the NIOSH recommended exposure limits (RELs).

Occupational exposure values are concentrations of airborne substances to which it is believed nearly all employees may be exposed throughout their working lifetime without suffering adverse health effects. These values are used as guidelines for evaluating exposures. Generally, safety factors are incorporated into the values, but values are not fine lines between safe and unsafe labels of exposure. Each individual exposure case must be evaluated based on several factors: the airborne concentration and its consistency, the material's warning properties and its acute/chronic health implications, individual susceptibility, and the significance of other potential exposure routes.

Most values represent TWA airborne concentrations for an 8–12-h workday, 40-h workweek. Limited exposures greater than the value are acceptable for most materials as long as the 8-h TWA exposure does not exceed the value. For some materials excursion values have also been established if the material could cause adverse effects from brief exposures to elevated concentrations. Frequency or duration limits are not established for the excursion values; rather, professional judgment should be used to assess the health implications of each individual exposure situation. A few materials have ceiling value limits, which indicate that the exposures should never exceed the ceiling value, even for brief periods. A "C" notation next to the value indicates the ceiling value.

Occupational exposure values are usually expressed in parts per million, volume by volume, or in milligrams of contaminant per cubic meter of air.

A designation of "A1" or "A2" indicates that the material is a confirmed human carcinogen (A1) or suspected human carcinogen (A2). A "skin" or "S" notation next to the value indicates that the material may be absorbed through the skin and may cause effects other than at the point of contact with the skin. For these materials skin contact may contribute to the overall exposure; therefore, inhalation exposure alone may not adequately characterize total exposure.

11.4 EXAMPLES OF CHEMICALS THAT REQUIRE MONITORING

11.4.1 Carbon Monoxide (CO)

CO is a colorless, odorless gas generated by the combustion of common fuels with an insufficient supply of air or where combustion is incomplete. It is often released by accident or the improper maintenance or adjustment of burners or flues in confined spaces and by internal combustion engines. Called "the silent killer," CO poisoning may occur suddenly.

11.4.2 Hydrogen Sulfide (H_2S)

H_2S, a colorless gas, initially smells like rotten eggs. However, the odor cannot be taken as a warning because sensitivity to smell disappears quickly after breathing only a small quantity of the gas. H_2S is flammable and explosive in high concentrations.

Sudden poisoning may cause unconsciousness and respiratory arrest. In less sudden poisoning symptoms are nausea, stomach distress, eye irritation, belching, coughing, headache, and blistering of lips.

11.4.3 Sulfur Dioxide (SO₂)

The combustion of sulfur or compounds containing sulfur produces SO_2, a pungent, irritating gas. Severe exposures may result from loading and unloading tank cars or cylinders, from rupturing or leaking pipes or tubing, and fumigation.

11.4.4 Ammonia (NH₃)

NH_3 is a strong irritant that can produce sudden death from bronchial spasms. Small concentrations that do not produce severe irritation are rapidly passed through the respiratory tract and metabolized so that they no longer act as ammonia. Ammonia can be explosive if the contents of a tank or refrigeration system are exposed to an open flame.

11.4.5 Benzene

Often the contaminant of greatest concern on a petroleum spill or remediation site is benzene with a PEL of 1 ppm. Benzene (C_6H_6) is a common component of gasoline and petroleum products, especially the higher-octane gasolines. Benzene is a colorless to light-yellow liquid with an aromatic odor. Exposure can cause symptoms of dizziness, light-headedness, headaches, and vomiting. High exposures may cause convulsions and coma and irregular heartbeat. Repeated exposure can damage the blood-forming organs, causing aplastic anemia. Long-term exposure can cause drying and scaling of the skin. Benzene is an A1 carcinogen proven to cause leukemia in humans.

Benzene released into the soil is subject to rapid volatilization near the surface. The benzene that does not evaporate will be highly to very highly mobile in the soil and may leach to groundwater. Benzene is uniformly distributed 1–10 cm through the soil and has a half-life of 7.2 to 38.4 days. If benzene is released into water, rapid volatilization may occur. Benzene will not adsorb to particulates. Biodegradation may occur. Benzene released to the atmosphere will exist in the vapor phase. Benzene is fairly soluble in water and is removed from the atmosphere by rain.

The first step if soil or water contamination by petroleum products is suspected is to assume that exposure during sampling will occur. Samplers will don HEPA-OV cartridge-equipped air-purifying respirators (APRs) and take soil samples. In lieu of laboratory analysis immunoassay field methods may be used.

When sustained benzene PID readings exceed 5 ppm, work must cease, and the EZ allowed to ventilate. Retesting and assigning of respirator protection may commence after a 30-min ventilation interval.

11.4.6 Hydrogen Cyanide or Hydrocyanic Acid (HCN)

HCN is an extremely rapid poison that interferes with the respiratory system and causes chemical asphyxia. Liquid HCN is an eye and skin irritant.

11.4.7 Lead

Lead is a heavy, soft gray metal. Lead exposure can cause a variety of health problems. The earliest symptoms may be tiredness, trouble sleeping, stomach problems, constipation, headaches, irritability, and depression. Higher levels may cause aching and weakness in the arms and legs, trouble concentrating and remembering things, and a low blood count (anemia). Lead exposure increases the risk of high blood pressure.

Repeated exposure can result in the buildup of lead in the body. This lead buildup is partially deposited in the bones. When referring to the amount of lead in the bones, the term *body burden* is often used. Body burden implies that the body is storing lead rather than excreting the lead through waste products or carrying the lead in the blood.

Because this lead is not being excreted in urine or carried in the blood, urine and blood samples will not reveal the total lead present in the body. Blood samples are an indication of lead exposure for approximately 2–4 weeks after the exposure incident. Then, as the body begins to deposit lead in the bone, blood samples become a less accurate indication of lead exposure.

The lead in bone may be released from the bone tissue when certain processes within the body occur. One of these processes occurs when the body begins to use the calcium stored in bone as a substitute for calcium lacking in the diet. When calcium is removed from bone, the lead held in the bone tissue also begins to enter the bloodstream. This process is one of the reasons why women of childbearing age are cautioned to avoid exposure to lead. Lead is a probable terratogen, which means that a developing fetus can be severely injured by exposure to lead.

Lead can cause serious permanent kidney or brain damage when exposures are high. Lead exposure can occur by inhalation or ingestion.

Lead if released or deposited in the soil will be retained in the upper 2–5 cm of soil, especially in soils with at least 5% organic matter or a pH of 5 or above. Leaching is not a significant process under most circumstances. Lead enters water from runoff or wastewater. Lead is effectively removed from the water column to the sediment by adsorption to organic matter and clay minerals. When released to the atmosphere, lead will generally occur as a dust or becomes adsorbed to particulate matter.

When lead dust levels reach 5 mg/m^3 (100 times 0.05 mg/m^3), air respirators must be worn. In all areas where visible dust is present and there is soil staining or other obvious signs of contamination (drum fragments, intact drums, chemical containers, buried treated wood), suspect lead contamination. Begin testing for both lead and PAH soil-adsorbed components using 2 l/min flow rate through filter-loaded cassettes. Filter analyticals for both lead and PAHs will be requested of the testing laboratory.

The OSHA PEL is 50 μg/m^3 (0.050 mg/m^3); the AL is 30 μg/m^3 (0.030 mg/m^3). On-site work may expose workers above the PEL. Biological monitoring of exposure is necessary if the airborne concentration exceeds 30 μg/m^3 (0.030 mg/m^3) for 30 days in 12 consecutive months.

On-site contaminant concentrations could exceed inhalation exposure maximum limits for lead. Consequently, HEPA cartridge-equipped APRs will be required for all personnel in any lead contaminant area during sampling activities. Air monitoring will be performed to assess the degree of exposure to lead particulates during on-site investigative work and to confirm the adequacy of the level of PPE being used.

Employee exposure is the exposure that would occur if the employee was not using a respirator. Full shift (for at least 7 continuous hours) personal samples, including at least one sample for each shift for each job classification in each work area, will be conducted in areas where lead contaminated soil is expected. Full shift personal samples will be representative of the monitored employee's regular, daily exposure to lead. Monitoring for the initial determination may be limited to a representative sample of the employees who the employer reasonably believes are exposed to the greatest airborne concentrations of lead in the workplace.

11.4.8 Flammable Chemicals

All flammable chemicals should be stored and used away from ignition sources such as open flames, cigarettes, and sparking tools. All vessels containing flammable chemicals will be grounded in accordance with OSHA and NFPA regulations and codes. Appropriate fire-extinguishing material will be kept available for fire emergencies. Flammable chemicals also pose a toxicological risk because the burn event may create new chemical formulations on a site. These chemicals may be spread either as out-gassing from the fire or via adsorption to fire smoke particulates and subsequent dispersion.

Flammable and combustible chemicals are defined by NFPA as materials that will rapidly or completely vaporize at atmospheric pressure and normal ambient temperature or that are readily dispersed in air and burn readily.

This following chart describes the various levels of flammable materials:

4   Gases
     Cryogenic materials
     - Any liquid or gaseous material that is a liquid while under pressure and has a flash point below 73°F (22.8°C) and a boiling point below 100°F (37.8°C) (Class IA flammable liquids)
     - Materials that on account of their physical form or environmental conditions can form explosive mixtures with air and that are readily dispersed in air, such as dusts of combustible solids and mists of flammable or combustible liquid droplets

3   Liquids and solids can be ignited under almost all ambient temperature conditions. Materials in this classification produce hazardous atmospheres with air under almost all ambient temperatures or, though unaffected by ambient temperatures, are readily ignited under almost all conditions. This classification should include the following:
     - Liquids having a flash point below 73°F (22.8°C) and having a boiling point at or above 100°F (37.8°C)
     - Liquids having a flash point at or above 73°F (22.8°C) and below 100°F (37.8°C) (Class IB and Class IC flammable liquids)
     - Solid materials in the form of coarse dusts that may burn rapidly, but generally do not form explosive atmospheres with air
     - Solid materials in a fibrous or shredded form that may burn rapidly and create flash fire hazards, such as cotton, sisal, and hemp
     - Materials that burn with extreme rapidity, usually by reason of self-contained oxygen (e.g., dry nitrocellulose and many organic peroxides)
     - Materials that ignite spontaneously when exposed to air

2   Materials that must be moderately heated or exposed to relatively high ambient temperatures before ignition can occur. Materials in this classification would not under normal conditions form hazardous atmospheres with air, but under high ambient temperatures or under moderate heating may release vapor in sufficient quantities to produce hazardous atmospheres with air. This classification should include the following:
     - Liquids having a flash point above 100°F (37.8°C), but not exceeding 200°F (93.4°F)
     - Solids and semisolids that readily give off flammable vapors

1   Materials that must be preheated before ignition can occur. Materials in this classification require considerable preheating, under all ambient temperature conditions, before ignition and combustion can occur. This classification should include the following:
     - Materials that will burn in air when exposed to a temperature of 1500°F (815.5°C) for a period of 5 min or less

- Liquids, solids, and semisolids having a flash point above 200°F (93.4°C)
  This classification includes most ordinary combustible materials.

0   Materials that will not burn. This classification should include any material that will not burn in air when exposed to a temperature of 1500°F (815.5°C) for a period of 5 min.

11.4.9 Reactive Hazards—Oxidizers

Oxidizers are chemicals that create a persistent fire when mixed with a flammable or combustible material. All oxidizer chemicals should be segregated from all flammable and combustible materials, including solvents, cleaners, paints, rags, paper, and wood. Personnel handling oxidizers will wear proper PPE and equipment.

Acid gases may be oxidizers in some situations and should be treated as such during monitoring. Remember the term *oxidation* means a chemical's ability to take electrons from another molecule. Acids, due to their positive valence potential, have the ability to take electrons from other chemicals. Thus the effects of oxidation may occur without what we traditionally think of as oxygen-bearing compounds being present.

11.4.10 Paint

Paint and painting supplies often contain a variety of hazardous substances, such as flammable solvents and toxic ingredients. Organic paints and paint thinners often contain flammable solvents that must be managed as other flammable chemicals. Aerosol sprays and epoxy resins sometimes contain toxic substances, including toluene diisocyanatates, and must therefore be scrutinized when the paint is initially purchased. Respiratory protection and/or adequate ventilation must always be used when working with paints.

11.4.11 Cleaning Supplies

Everyday common cleaning supplies must not be overlooked in a "right-to-know" compliance program. Corrosives and toxics are often used as ingredients in cleaning supplies.

11.4.12 Compressed Gases

Compressed gases must be managed to prevent accidental damage to the cylinder or the uncontrolled release of its gaseous contents.

- Damaged cylinders can become "unguided missiles."
- Uncontrolled releases of compressed gases could lead to asphyxiation.
- Stationary cylinders should be secured to walls or benches and should not be moved without a valve protector in place.

11.5 CONFINED SPACE MONITORING

Entry into any confined space is prohibited until its atmosphere has been tested from the outside. If entry is authorized, entry conditions will be continuously monitored in the areas where authorized entrants are working. The atmosphere within the space will be monitored using an O_2/CGI monitor equipped with CO toxin sensors.

Assume that every confined space has an unknown, hazardous atmosphere. Under no circumstances should anyone ever enter or even stick his or her head into a confined space for a "quick look." Such an action constitutes entry into the confined space and can expose the entrant to hazardous and possibly deadly atmospheres.

- The oxygen level must be determined first because most CGIs are oxygen dependent and will not provide reliable readings in an oxygen deficient atmosphere.
- Equipment for continuous monitoring of gases and vapors will be explosion proof and equipped with an audible alarm or danger-signaling device that will alert employees when a hazardous condition develops.
- Instruments used for testing the atmosphere in a confined space will be selected for their functional ability to measure hazardous concentrations.
- Instruments will be calibrated in accordance with the manufacturer's instructions. Each calibration will be recorded, filed, and available for inspection for 1 year after the last calibration date.

11.5.1 Entry Permits

Ongoing monitoring of the atmosphere will be performed in accordance with a confined space entry permit. When the atmospheric concentration of any substance cannot be kept within tolerance levels (i.e., PELs), then the employee will wear an approved respirator or leave the permit space.

The entry permit is revoked when the direct reading instrument being used or some other circumstance indicates that conditions in the space are no longer acceptable for entry. When an entry permit has been revoked because unacceptable conditions have arisen in a permit space, subsequent entry may not be made by special permit until the space is reevaluated by the entry supervisor.

11.5.2 Bump Testing

Each day before monitoring a space, the instrument must be bump tested. The purpose of bump testing is to assure the readings on the instrument display are within the limits stated on the calibration cylinder. Bump testing is accomplished by

- Turning the instrument on
- Allowing the instrument to warm up for at least 10 min
- Passing a known concentration of calibrated gas through the pump module across the sensors

Instruments will only be used by employees who have been trained in the proper operation, use, limitations, and calibration of the monitoring instrument.

11.5.3 Monitoring for LEL and O_2 Levels

In any confined space classified as a class II or class III hazardous location according to the National Electrical Code, Article 500, Sections 6 and 7, a fire watch will be established as part of the entry procedure. In such areas surface dust and fibers will be removed, and no work initiated until the airborne particulate level is below 10% of the LEL for the material.

When combustible dusts or ignitable fibers/filings are present, all equipment and ventilation systems used in the confined space will comply with Articles 502 and 503 of the National Electrical Code.

11.5.4 Isolation

If isolation of the space is not feasible because the space is large or is part of a continuous system (such as a sewer), preentry testing will be performed to the extent possible before entry is authorized. Any necessary additional tests will be selected and performed to the satisfaction of the entry supervisor (i.e., substance-specific detectors should be used whenever actual or potential contaminants have been identified).

11.5.5 Confined Space—Cautionary Statements

If possible, do not open the entry portal to the confined space and draw the sample through a small entry port leading into the confined space. Sudden changes in atmospheric composition within the confined space could cause violent reactions, or dilute the contaminants in the confined space, giving a false low initial gas concentration.

Comprehensive testing should be conducted in various locations within the work area. It is important to understand that some gases or vapors are heavier than air (e.g., hydrogen sulfide) and some gases (e.g., methane) are lighter than air. Therefore, all areas (top, middle, bottom) of a confined space must be tested with properly calibrated testing instruments.

The results of atmospheric testing will have a direct impact on the selection of protective equipment necessary for the tasks in the confined area. These results may also dictate the duration of worker exposure to the environment of the space or whether an entry will be made at all.

11.5.6 Stratified Atmospheres

When monitoring for entries involving a descent into atmosphere that may be stratified,

- The atmosphere envelope should be tested a distance of approximately 4 ft in the direction of travel and to each side.
- The entrant's rate of progress should be slowed to accommodate the sampling speed and detector response of the sampling probe.

11.6 WELDING

The most significant hazard in the welding process is the generation of fumes and gases. The amount and type of fumes and gases involved will depend on the welding process, the base material, the filler material, and the shielding gas. The toxicity of the contaminants depends primarily on contaminant concentrations and the physiological responses of the human body. A number of potentially hazardous materials are used in fluxes, coatings, coverings, and filler metals. Some of these include beryllium, cadmium, cleaning compounds, fluorine compounds, lead, mercury, stainless steels, and zinc. The suppliers of these materials must determine if any hazard is associated with welding and cutting and provide warnings through tags, signs, etc. on boxes and containers. Employers also must follow the ventilation requirements specified in the standards for these materials.

Mechanical ventilation must be provided when welding or cutting is done on other metals. Mechanical local exhaust ventilation may be provided either by means of freely movable hoods or fixed enclosures with tops to provide sufficient ventilation.

11.6.1 Effects of Toxic Gases

Exposure to various toxic gases generated during welding processes may produce one or more of the following effects:

- Inflammation of the lungs (chemical pneumonitis)
- Pulmonary edema (swelling and accumulation of fluids)
- Emphysema (loss of elasticity of the lungs) (A very small percentage of emphysema is caused by occupational exposure.)
- Chronic bronchitis
- Asphyxiation

The major toxic gases associated with welding are classified as primary pulmonary and nonpulmonary. Cleaning compounds because of their properties often require special ventilation precautions following the manufacturer's instructions. Degreasing operations may involve chlorinated hydrocarbons; these liquids or vapors should be kept away from molten weld metal or the arc. Also keep them away from ultraviolet radiation from welding operations.

11.6.2 Ventilation

Local exhaust or general ventilating systems must be provided and arranged to keep the amount of toxic fumes, gases, or dust below the maximum allowable concentration as specified in OSHA's standard on air contaminants (29 CFR 1910.1000).

Natural ventilation is acceptable for welding, cutting, and related processes, where the necessary precautions are taken to keep the welder's breathing zone away from the plume and where sampling of the atmosphere shows that concentrations of contaminants are below unsafe levels. Natural ventilation often meets the conditions, where the necessary precautions are taken to keep the welder's breathing zone away from the plume and all of the following conditions are met:

- Space of more than 10,000 ft^3 (284 m^3) per welder is provided.
- Ceiling height is more than 16 ft (5 m).
- Welding is not done in a confined space.
- Welding space does not contain partitions, balconies, or other structural barriers that significantly obstruct cross-ventilation. Welding space refers to a building or an enclosed room in a building, not a welding booth or screened area that is used to provide protection from welding radiation.
- Materials covered above are not present as deliberate constituents.

The only way to assure that airborne contaminant levels are within the allowable limits, however, is to take air samples at the breathing zones of the personnel involved.

Mechanical ventilation includes local exhaust, local forced, and general area mechanical air movement. Local exhaust ventilation is preferred.

- Local exhaust ventilation means fixed or movable exhaust hoods placed as near as practicable to the work area to maintain a capture velocity sufficient that keeps airborne contaminants below unsafe limits.
- Local forced ventilation means a local air-moving system (such as a fan) placed so that it moves the air at right angles (90°) to the welder (across the welder's face). It should produce an approximate velocity of 100 ft/min (30 m/min) and be maintained for a distance of approximately 2 ft (600 mm) directly above the work area. Precautions must be taken to insure that contaminants are not dispersed to other work areas.

General mechanical ventilation may be necessary to maintain the general background level of airborne contaminants below the levels referred to or to prevent the accumulation of explosive gas mixture. Examples of general mechanical ventilation are roof exhaust fans, wall exhaust fans, and similar large area air movers. General mechanical ventilation is not usually as satisfactory for health hazard control as local mechanical ventilation. It is often helpful, however, when used in addition to local ventilation.

11.6.3 Ventilation in Confined Spaces during Welding

All welding and cutting operations carried out in confined spaces must be adequately ventilated to prevent the accumulation of toxic materials or possible oxygen deficiency. Oxygen should never be used for ventilation. When it is impossible to provide adequate ventilation, airline respirators or hose masks approved by the NIOSH must be used.

In areas immediately hazardous to life, hose masks with blowers or self-contained breathing equipment that has been approved by the NIOSH must be used. Where welding operations are being carried out in a confined space, and welders and helpers are provided with hose masks or hose masks with blowers or self-contained breathing equipment, an employee must be stationed outside the space to insure the safety of those working within.

11.6.4 Fume Avoidance

Welders and cutters must take precautions to avoid breathing the fume plume directly by adjusting the position of the work or the head or by ventilation that directs the plume away from the face.

11.6.5 Light Rays

Electric arcs and gas flames produce ultraviolet and infrared rays; continuous or repetitious ultraviolet exposure has a harmful effect on the eyes and skin. The usual ultraviolet effect is "sunburn" of the surface of the eye, which is painful and disabling, but temporary in most instances. However, permanent eye injury may result from looking directly into a very powerful arc without eye protection, due to the effect of visible and near infrared radiation. Ultraviolet radiation may also produce the same effects on the skin as a severe sunburn.

The production of ultraviolet radiation is high in gas-shielded arc welding. For example, a shield of argon gas around the arc doubles the intensity of the ultraviolet radiation, and, with the greater current densities required (particularly with a consumable electrode), the intensity may be 5–30 times as great as with nonshielded welding, such as covered electrode or gas-shielded metal arc welding.

11.6.6 Infrared Rays

Infrared radiation has the effect only of heating the tissue with which it comes in contact. If the heat is not enough to cause an ordinary thermal burn, there is no harm.

Whenever possible, arc-welding operations should be isolated so that other workers will not be exposed to either direct or reflected rays. Arc-welding stations for regular production work can be enclosed in booths if the size of the work area permits. The inside of the booth should be coated with a paint that is nonreflective to ultraviolet radiation and provided with portable flameproof screens similarly painted or with flameproof curtains. Booths should be designed to permit air circulation at the floor level and adequate exhaust ventilation.

11.6.7 Noise

In welding and cutting and the associated operation, noise levels may exceed the permissible limits. Personal hearing protective devices may be needed.

11.6.8 Gas Welding and Cutting

Mixtures of fuel gases and air or oxygen in gas welding and cutting must not be permitted, except immediately prior to consumption, to guard against explosions. Acetylene must not be generated, piped (except in approved cylinder manifolds), or used at a pressure higher than 15 $lb/in.^2$ gauge or 30 $lb/in.^2$ absolute. Only approved apparatus such as torches, regulators, or pressure-reducing valves, acetylene generators, and manifolds must be used. All portable cylinders used for storage and shipment of compressed gases must be constructed and maintained according to Department of Transportation regulations.

Risk Communication and Environmental Monitoring

When results must be communicated to the public, Chapter 12 provides proven methodologies for explaining the relevancy and adequacy of monitoring results.

This chapter includes regulations, responsibilities, and compliance requirements associated with air pollution emissions from stationary and mobile sources. The significant types and sources of air pollution emissions include the following:

- Particulates, SO_2, NO_x, CO, VOCs, and hazardous air pollutants from fuel burning at steam and hot water generation plants and boilers
- Particulates and toxic air emissions from the operation of hazardous waste; general waste; classified material; and medical, pathological, and/or infectious waste incinerators
- Particulates, CO, metals, and toxic air pollutant emissions from open burning and open detonation operations
- VOC vapor emissions from the operation of degreasers and other processes (paint stripping and metal finishing) that use solvents
- CO emissions from vehicles and equipment operated on the installation/installation facility
- Fugitive particulate emissions from training activities and construction/demolition operations

Most facilities have air emissions sources in one or more of these categories.

12.1 FEDERAL LEGISLATION

12.1.1 The Clean Air Act Amendments of 1990 (CAAA90)

Public Law (PL) 101-549 (42 U.S. Code [USC] 7401-7671q), known as CAAA90, is the current federal legislation regulating the prevention and control of air pollution. The act has seven major titles that address various aspects of the national air pollution control program:

- Title I describes air pollution control requirements for geographic areas in the U.S. with respect to the National Ambient Air Quality Standards (NAAQS).
- Title II deals mostly with revised tailpipe emission standards for motor vehicles. These requirements compel automobile manufacturers to improve design standards to limit CO, hydrocarbon, and NO_x emissions. Oxygenated gasoline is required in cities with the worst ozone and CO nonattainment. Reformulated gasoline and gasoline with reduced Reid vapor pressure is used in ozone non-attainment areas.
- Title III potentially contains the most costly requirement of CAAA90. Its major elements deal with hazardous air pollutants through control of routine emissions and contingency planning for accidental releases.
- Title IV addresses acid deposition control and applies only to commercial utilities that produce electricity for sale.
- Title V outlines the requirement of having states issue federally enforceable operating permits to major stationary sources. The permits are designed to enhance the ability of the EPA, state regulatory agencies, and private citizens to enforce the requirements of CAAA90. Permits will also be used to specify operation and control requirements for stationary sources.
- Title VI limits the emissions of chlorofluorocarbons (CFCs), halons, and other halogenated chemicals that contribute to the destruction of stratospheric ozone. These requirements closely follow the control strategies recommended in June 1990 by the second meeting of parties to the Montreal Protocol. Also, procurement of ozone-depleting substances is restricted by Department of Defense acquisition policies.
- Title VII describes civil and criminal penalties that may be imposed for the violation of new and existing air pollution control requirements. This title also gives authority to the EPA to issue field citations for many types of violations.

12.1.2 The Federal Water Pollution Control Act

The Federal Water Pollution Control Act, commonly known as the Clean Water Act (CWA), governs the control of water pollution in the U.S. The objective of this act is to restore and maintain the chemical, physical, and biological integrity of the nation's waters.

The CWA led to the promulgation of regulations concerning the incineration of sewage sludge. The implementing regulations for the control of emissions from the incineration of sewage sludge are found in 40 CFR 503.40 through 503.48.

12.1.3 Resource Conservation and Recovery Act (RCRA) of 1976

The RCRA is the federal law that governs the disposal of solid waste. Subtitle D of the RCRA, as last amended in November 1984, PL 98-616, 42 USC 6941-6949a, establishes federal standards and requirements for state and regional authorities respective to solid waste disposal. The objectives of this subtitle are to assist in developing and encouraging methods for the disposal of solid wastes that are environmentally sound and that maximize the utilization of valuable resources recoverable from solid waste. The objectives are to be achieved through federal technical and financial assistance to states and regional authorities for comprehensive planning (42 USC 6941).

12.1.4 State/Local Regulations

The primary mechanisms regulating air pollutant emissions are state air quality regulations. These regulations will normally follow the federal guidelines for state programs and will have many similar features. However, depending on the type and degree of air pollutant problems within a state/region, the individual regulations will vary. As an example, ozone problems are widespread in California; therefore, various local authorities in that state have stringent VOC emission requirements. The state of North Dakota has no such problem and, therefore, has fewer and less stringent VOC regulations.

A permit is normally required for new, expanded, or modified sources of air pollutants. There are federal, state, and local permits required for various sources. Large sources, and the installation/CW facility as a whole, may require a permit to operate. States review permit applications for the construction or operation of many sources. Open burning permits are typically handled locally.

Some state regulations apply directly to some installations/CW facilities and operations without requiring a permit. At a minimum state regulations should be reviewed for the following activities:

- Fugitive dust emissions
- Control of particulate emissions from the transportation of refuse or materials in open vehicles
- Certification requirements for boiler operators
- Emissions and emission control requirements for the operation of existing fossil fuel-fired steam generators
- Open burning
- Vehicle exhaust emissions testing
- Spray painting of vehicles, buildings, and/or furniture
- Certification of vehicles transporting VOC liquids
- Paving of roads and parking lots
- Toxic air pollutants
- Operation of cold cleaners, degreasers, and open-top vapor degreasers
- Vapor control requirements for fuel pumps

12.2 KEY COMPLIANCE REQUIREMENTS

12.2.1 Steam-Generating Units [greater than 29 MW (100 MBtu/h)]

Steam-generating units (with a capacity > 29 MW) that started construction or modification after June 19, 1984 are required to meet emissions limitations for particulates, SO_2, and NO_x. The limit that applies is dependent on the type of fuel being burned. Records of the amounts of fuel combusted each day are required (40 CFR 60.40b through 49b).

12.2.2 Steam-Generating Units [2.9 MW (10 MBtu/h) to 29 MW]

Steam-generating units (with a maximum design heat input capacity ≥ 2.9 MW but ≤ 29 MW) that started construction, modification, or reconstruction after June 3, 1989 are required to limit emissions of SO_2 and particulates. Emission rates must be monitored. Installations/CW facilities are required to submit excess emission reports for any calendar quarter in which the facility exceeds opacity limits. If an installation/CW facility does not exceed the limits in a given year, it is required to file semiannual reports affirming this fact. Installations/CW facilities required to meet SO_2 emission limits are also required to submit quarterly reports (40 CFR 60.40c through 60.48c).

12.2.3 Fuel-Burning Facilities

Fuel-burning facilities (with heat input > 73 MW [250 BTU/h]) constructed or modified after August 17, 1971 are required to limit emissions of particulates, SO_2, and NO_x. Monitoring of these pollutants and fuel analysis is also required (40 CFR 60.44 and 60.45).

12.2.4 Stationary Gas Turbines

Stationary gas turbines (with a heat input $\geq$ 10.7 gJ/h [10 MBtu/h]) that were constructed or modified after October 3, 1977 are required to limit the emissions of NO_x and SO_2. In addition to the emissions, the sulfur and nitrogen content of the fuel being fired must also be monitored (40 CFR 60.330 through 60.335).

12.2.5 Municipal Waste Combustor

Municipal waste combustors (with a capacity > 225 metric tons [250 t] per day) that started construction or modification after December 20, 1989 are required to limit the amounts of dioxin/furan, SO_2, hydrogen chloride, CO, and NO_x emitted. The chief municipal waste combustor operator and shift supervisors are required to be certified to operate the combustor, and there must be an operating manual that is updated yearly (40 CFR 60.50a through 60.58a).

12.2.6 Incinerators

Incinerators (with a charging rate of $\geq$ 45 metric tons/day [50 t/day]) that started construction or modification after August 17, 1971 are required to meet emissions limitations for particulates. Additionally, they must maintain records of daily charging rates and hours of operation (40 CFR 60.50 through 60.54).

12.2.6.1 Sewage Sludge Incinerators

Sewage sludge incinerators (that combust > 1000 kg/day [2205 lb/day]) that were constructed or modified after June 11, 1973 are required to limit their emissions of particulates. Monitoring devices are required, depending on what type of incinerator the installation/CW facility operates. Semiannual reports are required (40 CFR 60.150 through 60.156).

12.2.6.2 Beryllium Incinerators

Incinerators for beryllium-containing waste, beryllium, beryllium oxide, or beryllium alloys cannot emit into the atmosphere more than 10 g (0.32 oz) of beryllium over a 24-h period. Records of emissions tests are to be kept for 2 years (40 CFR 61.30 through 61.34).

12.2.6.3 Incineration of Sewage Sludge

Installations/facilities with incinerators that fire sewage sludge must meet specific emissions standards for beryllium emissions, mercury emissions, and hydrocarbons. The incinerators must have continuous monitoring devices for hydrocarbons and oxygen in the exit gas, with continuous monitoring for combustion temperature as specified by the permitting authority. Assorted reports must be submitted and records kept (40 CFR 503.40 through 503.48).

12.2.7 Gasoline Dispensing

Leaded gasoline should not be introduced into any motor vehicle that is labeled "Unleaded gasoline only" or that is equipped with a gasoline tank filler inlet designed for unleaded gasoline. Fuel pumps are required to display signs stating the type of fuel in each pump and that only unleaded gas can be introduced into labeled vehicles. The nozzles of the pumps must be properly sized. Depending on whether the oxygenated gas is still in the control period, or the area has an oxygenated gasoline program with a credit program, pumps dispensing oxygenated gasoline must have required labels.

Since 1992, during high ozone seasons and regulatory control periods, gasoline cannot be sold, offered for sale, imported, dispensed, supplied, or transported that exceeds the Reid vapor pressure standards. No diesel fuel is to be distributed, transported, offered for sale, or dispensed for use in motor vehicles unless it is free of the dye 1,4-dialkylamino-antraquinone and has an octane index of at least 40 or a maximum aromatic content of 35 volume percent and a sulfur percentage of less than 0.05% [40 CFR 80.22(a), 80.22(d), 80.22(e), 80.24(a)(1), 80.27(a)(2), 80.35, 80.80(d), and 80.29(a)].

12.2.8 Rotogravure Printing Presses

Rotogravure printing presses, except for proof presses, that started construction or modification after October 28, 1980 are required to ensure that gases are not being discharged with VOCs equal to more than 16% of the total mass of VOC solvent and water used at that press during any one performance averaging period (40 CFR 60.430 through 60.435).

12.2.9 Fugitive Emissions

The emission of volatile hazardous air pollutants (VHAPs), vinyl chloride, and benzene is to be managed, monitored, and controlled according to specific requirements. These requirements include taking certain actions when a leak is detected, ensuring that certain records are maintained, ensuring that pumps and compressors meet certain standards, and that pressure relief devices in gas/vapor service have no detectable emissions except during pressure releases. Valves and lines in VHAP service must be monitored monthly and repairs done within 15 days of leak detection. Systems and devices used to control VHAP emissions must recover vapors with a 95% efficiency or greater. Enclosed combustion devices are to be designed and operated to reduce VHAP and benzene emissions, and closed-vent systems will have no detectable emissions (40 CFR 61.240 through 61.242-10, 61.246, and 61.247).

12.2.10 Sulfuric and Nitric Acid Plants

Sulfuric and nitric acid plants are required to limit their emissions and install continuous monitoring systems (40 CFR 60.70 through 60.85).

12.2.11 CFCs and Halons

To protect the ozone layer, no person repairing or servicing motor vehicles for payment can service a motor vehicle air-conditioner (MVAC) in any way that affects the refrigerant unless they have been trained and certified and are using approved equipment. Additionally, persons who maintain, service, or repair appliances, except MVACs, and persons who dispose of appliances, except for small appliances, room air conditioners, MVACs, and MVAC-like appliances are required to be certified through an approved technician certification program.

As of November 15, 1992 no class I or class II substances suitable for use in motor vehicles as a refrigerant can be sold or distributed in any container that is less than 20 lb (9 kg) to any person unless that person is trained and certified. Installations/facilities that sell class I or class II substances suitable for use as a refrigerant in containers of less than 20 lb (9 kg) are required to display a sign couched in specific language. Servicing appliances containing CFCs and halons is required to be done in a manner to prevent emissions [40 CFR 82.34(a), 82.34(b), 82.42(a) through 82.42(c), and 80.150 through 80.166].

12.2.12 Degreasing Operations

Batch cold-cleaning machines, batch vapor-cleaning machines, and in-line-cleaning machines must have tightly fitting covers and assorted emission control devices to prevent excess emissions. Operators of these types of units are also required to submit notifications, operating reports, exceedance reports, and solvent use reports. These regulations specifically apply to each individual batch vapor, in-line vapor, in-line cold, and batch cold solvent cleaning machine that uses any solvent containing methylene chloride, perchloroethylene, trichloroethylene, 1,1,1-trichloroethane, carbon tetrachloride, chloroform, or any combination of these halogenated Hazardous Air Pollutants (HAP) solvents, in total concentration greater than 5% by weight, as a cleaning and/or drying agent. Special sampling pumps are available to monitor grease dispersion (Figure 12.1).

12.3 KEY COMPLIANCE DEFINITIONS

(**Note:** The definitions provided are specific to the regulations cited above. Other regulations may define the terms differently.)

Across Rod Solvent-Cleaning Machine: a batch solvent-cleaning machine in which parts baskets are suspended from "cross-rods" as they are moved through the machine. Parts are loaded semicontinuously and enter and exit the machine from a single portal (40 CFR 63.431).

Air Blanket: the layer of air inside the solvent-cleaning machine freeboard located above the solvent/air interface. The centerline of the air blanket is equidistant between the sides of the machine (40 CFR 63.461).

Figure 12.1 The filter funnel holds a 37-mm filter in place during solvent extractions over a length of time as specified in ASTM Provisional Standard Test Method for Metalworking Fluids (ASTM PS 42-97). (SKC)

Air Pollution Control Device: one or more processes used to treat the exit gas from a sewage sludge incinerator stack [40 CFR 503.41(a)].

Ancillary Equipment: the equipment used in a dry-cleaning system including, but not limited to, emission control devices, pumps, filters, muck cookers, stills, solvent tanks, solvent containers, water separators, exhaust dampers, diverter valves, interconnecting piping, hoses, and ducts (40 CFR 63.321).

Annual Capacity Factor: the ratio between the actual heat input to a steam-generating unit from an individual fuel or combustion of fuels during a period of 12 consecutive calendar months and the potential heat input to the steam-generating unit from all fuels, if the steam-generating unit had been operated for 8700 h during that 12-month period at the maximum design heat input capacity (40 CFR 60.41c).

Appliance: any device that contains and uses a class I or class II substance as a refrigerant and that is used for household or commercial purposes, including any air conditioner, refrigerator, chiller, or freezer (40 CFR 82.152).

Apprentice: any person who is currently registered as an apprentice in service, maintenance, repair, or disposal of appliances with the U.S. Department of Labor's Bureau of Apprenticeship and Training (or a state apprenticeship council recognized by the Bureau of Apprenticeship and Training). If more than 2 years have elapsed since the person first registered as an apprentice, the person is not recognized as an apprentice (40 CFR 82.152).

Approved Equipment Testing Organization: any organization that has applied for and received approval from the administrator pursuant to 40 CFR 82.160 (40 CFR 82.152).

Area Source: any perchloroethylene dry-cleaning facility that is not a major source (40 CFR 63.321).

Articles: clothing, garments, textiles, fabrics, leather goods, and the like that are dry-cleaned (40 CFR 63.321).

Automated Parts Handling System: a mechanical device that carries all parts and parts baskets at a controlled speed from the initial loading of soiled or wet parts through the removal of the cleaned or dried parts. Automated parts handling systems include, but are not limited to, hoists and conveyors (40 CFR 63.461).

Auxiliary Fuel: fuel used to augment the fuel value of sewage sludge. This includes, but is not limited to, natural gas, fuel oil, coal, gas generated during anaerobic digestion of sewage sludge, and municipal solid waste (not to exceed 30% of the dry weight of sewage sludge and auxiliary fuel together) [40 CFR 503.41(b)].

Batch-Cleaning Machine: a solvent-cleaning machine in which individual parts or a set of parts move through the entire cleaning cycle before new parts are introduced into the machine. An open top, vapor-cleaning machine is a type of batch-cleaning machine. A solvent-cleaning machine, such as a ferris wheel or a cross-rod degreaser, that cleans multiple batch loads simultaneously and is manually loaded is a batch-cleaning machine (40 CFR 63.461).

Batch Municipal Waste Combustor: an incinerator that operates by forcefully projecting a curtain of air across an open chamber or pit in which burning occurs. Incinerators of this type can be constructed above or below ground and with or without refractory walls and floor (40 CFR 60.51b).

Benzene Service: a piece of equipment that either contains or contacts a fluid (liquid or gas) that is at least 10% benzene by weight (40 CFR 61.111).

Bulk Gasoline Terminal: any gasoline facility that receives gasoline by pipeline, ship, or barge and has a throughput >75,000 l/day (19,998 gal/day) (40 CFR 60.501).

Carbon Adsorber: a bed of activated carbon into which an air–perchloroethylene gas–vapor stream is routed and which adsorbs the perchloroethylene on the carbon (40 CFR 63.321). A bed of activated carbon into which an air solvent gas vapor stream is routed and that adsorbs the solvent on the carbon (40 CFR 63.461).

Cartridge Filter: a discrete filter unit containing both filter paper and activated carbon that traps and removes contaminants from petroleum solvent, together with the piping and ductwork used in installing this device (40 CFR 60.621).

Certified Refrigerant Recovery or Recycling Equipment: equipment certified by an approved equipment testing organization to meet the standards in 40 CFR 82.158(b) or (d), equipment certified pursuant to 40 CFR 82.36(a), or equipment manufactured before November 15, 1993 that meets the standards in 40 CFR 82.158(c), (e), or (g) (40 CFR 82.152).

Clean Liquid Solvent: fresh, unused solvent, recycled solvent, or used solvent that has been cleaned of soils (e.g., skimmed of oils or sludge and strained of metal chips) (40 CFR 63.461).

Cleaning Capacity: for a cleaning machine without a solvent/air interface, the maximum volume of parts that can be cleaned at one time. In most cases the cleaning capacity is equal to the volume (length times width times height) of the cleaning chamber (40 CFR 63.461).

Closed-Vent System: a system that is not open to the atmosphere and is composed of piping, connections, and, if necessary, flow-inducing devices that transport gas or vapor from a piece or pieces of equipment to a control device (40 CFR 61.241).

Coal Refuse: any waste products of coal mining, cleaning, and coal preparation operations (e.g., cull, gob) containing coal, matrix material, clay, and other organic and inorganic material (40 CFR 60.41a).

Cofired Combustor: a unit combusting municipal solid waste with nonmunicipal solid waste fuel (e.g., coal, industrial process waste) and subject to a federally enforceable permit limiting the unit to combusting a fuel feed stream, 30% or less of which is composed of municipal solid waste as measured on a calendar quarter basis (40 CFR 60.51a and 60.51b).

Cogeneration Steam-Generating Unit: a steam-generating unit that simultaneously produces both electrical (or mechanical) and thermal energy from the same primary energy source (40 CFR 60.41c).

Coin-Operated Dry-Cleaning Machine: a dry-cleaning machine that is operated by the customer (i.e., the customer places articles into the machine, turns the machine on, and removes articles from the machine) (40 CFR 63.321).

Cold-Cleaning Machine: any device or piece of equipment that contains and/or uses liquid solvent into which parts are placed to remove soils from the surface of the parts or to dry the parts. Cleaning machines that contain and use heated, non-boiling, solvent to clean the parts are classified as cold-cleaning machines (40 CFR 63.461).

Colorimetric Detector Tube: a glass tube (sealed prior to use) containing material impregnated with a chemical that is sensitive to perchloroethylene and is designed to measure the concentration of perchloroethylene in the air (40 CFR 63.321).

Combustion Research: the experimental firing of any fuel or combination of fuels in a steam-generating unit for the purpose of conducting research and development of more efficient combustion or more effective prevention or control of air pollution emissions from combustion, provided that, during these periods of research and development, the heat generated is not used for any purpose other than pre-heating combustion air for use by that steam-generating unit (i.e., the heat generated is released to the atmosphere without being used for space heating, process heating, driving pumps, preheating combustion air for other units, generating electricity, or any other purpose) (40 CFR 60.41c).

Commercial Refrigeration: refrigeration appliances utilized in the retail food and cold storage warehouse sectors. Retail food appliances include the refrigeration equipment found in supermarkets, convenience stores, restaurants, and other food service establishments. Cold storage appliances include the equipment used to store meat, produce, dairy products, and other perishable goods. All of the equipment contains large refrigerant charges, typically over 75 lb (34 kg) (40 CFR 82.152).

Commercial/Retail Waste: material discarded by stores, offices, restaurants, warehouses, nonmanufacturing activities at industrial facilities, and other similar establishments or facilities (40 CFR 60.51a).

Continuous Cleaning Machine: see in-line-cleaning machine.

Continuous Emissions Monitoring System (CEMS): a monitoring system for continuously measuring the emissions of a pollutant from an affected facility (40 CFR 60.51a and 60.51b).

Control Efficiency: the mass of a pollutant in the sewage sludge fed to an incinerator minus the mass of that pollutant in the exit gas from the incinerator stack divided by the mass of the pollutant in the sewage sludge fed to the incinerator [40 CFR 503.41(c)].

Critical Component: a component without which industrial process refrigeration equipment will not function, will be unsafe in its intended environment, and/or will be subject to failures that would render the industrial process served by the refrigeration appliance unsafe (40 CFR 82.152).

Custom-Built: specialized equipment or any of its critical components cannot be purchased and/or installed without being uniquely designed, fabricated, and/or assembled to satisfy a specific set of industrial process conditions (40 CFR 82.152).

Designated Volatility Nonattainment Area: any area designated as being in nonattainment with the NAAQS for ozone pursuant to rule making under Section 107(d)(4)(A)(ii) of CAAA90 (40 CFR 80.2).

Designated Volatility Attainment Area: an area not designated as being in nonattainment with the NAAQS for ozone (40 CFR 80.2).

Diesel Fuel: any fuel sold in any state and suitable for use in diesel motor vehicles and diesel motor vehicle engines that is commonly or commercially known or sold as diesel fuel (40 CFR 80.2).

Dispersion Factor: the ratio of the increase in the ground level ambient air concentrations for a pollutant at or beyond the property line of the site where the sewage sludge incinerator is located to the mass emission rate for the pollutant from the incinerator stack [40 CFR 503.41(d)].

Disposal: the process leading to and including the discharge, deposit, dumping, or placing of any discarded appliance into or on any land or water; the disassembly of any appliance for discharge, deposit, dumping, or placing of its discarded component parts into or on any land or water; and the disassembly of an appliance for reuse of its component parts (40 CFR 82.152).

Diverter Valve: a flow control device that prevents room air from passing through a refrigerated condenser when the door of the dry-cleaning machine is open (40 CFR 63.321).

Dry-Cleaning Cycle: the washing and drying of articles in a dry-to-dry machine or transfer machine system (40 CFR 63.321).

Dry-Cleaning Facility: an establishment with one or more dry-cleaning systems (40 CFR 63.321).

Dry-Cleaning Machine: a dry-to-dry machine or each machine of a transfer machine system (40 CFR 63.321).

Dry-Cleaning Machine Drum: the perforated container inside the dry-cleaning machine that holds the articles during dry cleaning (40 CFR 63.321).

Dry-Cleaning System: a dry-to-dry machine and its ancillary equipment or a transfer machine system and its ancillary equipment (40 CFR 63.321).

Dryer: a machine used to remove petroleum solvent from articles of clothing or other textile or leather goods, after washing and removing excess petroleum solvent, and the piping and ductwork used in the installation of this device (40 CFR 60.621). A machine used to remove perchloroethylene from articles by tumbling them in a heated airstream (40 CFR 63.321).

Dry-to-Dry Machine: a one machine, dry-cleaning operation in which washing and drying are performed in the same machine (40 CFR 63.321).

Duct Burner: a device that combusts fuel and is placed in the exhaust duct from another source (such as a stationary gas turbine, internal combustion engine, kiln, etc.) to allow the firing of additional fuel to heat the exhaust gases before the exhaust gases enter a steam-generating unit (40 CFR 60.41c).

Dwells: the technique of holding parts within the freeboard area, but above the vapor zone of the solvent-cleaning machine. Dwell occurs after cleaning to allow solvent to drain from the parts or parts baskets back into the solvent-cleaning machine (40 CFR 63.461).

Emerging Technology: any SO_2 control system that is not defined as a conventional technology and for which the owner or operator of the affected facility has received approval from the administrator to operate as an emerging technology (40 CFR 60.41c).

Exhaust Damper: a flow control device that prevents the air–perchloroethylene gas–vapor stream from exiting the dry-cleaning machine into a carbon adsorber before room air is drawn into the dry-cleaning machine (40 CFR 63.321).

Existing: in relation to perchloroethylene dry cleaners, construction or reconstruction commenced before December 9, 1991 (40 CFR 63.321). Any solvent-cleaning machine whose construction or reconstruction commenced on or before November 29, 1993, but did not meet the definition of a solvent-cleaning machine on December 2, 1994 because it did not use halogenated HAP solvent liquid or vapor covered under this subpart to remove soils, becomes an existing source when it commences to use such liquid or vapor. A solvent-cleaning machine moved within a contiguous facility or to another facility under the same owner-ship constitutes an existing machine (40 CFR 63.461).

Federally Enforceable: all limitations and conditions enforceable by the administrator, including those requirements developed pursuant to 40 CFR 60 and 61, requirements within any applicable state implementation plan, and any permit requirements established pursuant to 40 CFR 52.21 or under 40 CFR 51.18 and 51.24 (40 CFR 60.41b).

Fluidized Bed Incinerator: an enclosed device in which organic matter and inorganic matter in sewage sludge are combusted in a bed of particles in the combustion chamber gas [40 CFR 503.41(e)].

Follow-up Verification Test: checking the repairs within 30 days of an appliance's return to normal operating characteristics and conditions. Follow-up verification tests for appliances from which the refrigerant charge has been evacuated means a test conducted after the appliance or portion of the appliance has resumed operating at normal operating characteristics and conditions of temperature and pressure, except in cases where sound professional judgment dictates that these tests will be more valid if they are performed prior to the return to normal operating characteristics and conditions. A follow-up verification test with respect to repairs conducted without evacuation of the refrigerant charge means a reverification test conducted after the initial verification test and usually within 30 days of the return to normal operating conditions. If an appliance is not evacuated, it is only necessary to conclude that any required changes in pressure, temperature, or other conditions returned the appliance to normal operating characteristics and conditions (40 CFR 82.152).

Fossil Fuel: natural gas, petroleum, coal, and any form of solid, liquid, or gaseous fuel derived from such materials for the purpose of creating useful heat (40 CFR 60.41a).

Freeboard Area: for a batch-cleaning machine this is the area within the solvent-cleaning machine that extends from the solvent/air interface to the top of the machine; for an in-line-cleaning machine, it is the area within the solvent-cleaning machine that extends from the solvent/air interface to the bottom of the entrance or exit opening, whichever is lower (40 CFR 63.461).

Freeboard Ratio: the ratio of the solvent-cleaning machine freeboard height to the smaller interior dimensions (length, width, or diameter) of the solvent-cleaning machine (40 CFR 63.461).

Fuel Pretreatment: a process that removes a portion of the sulfur in a fuel before combustion in a steam-generating unit (40 CFR 60.41c).

Fugitive Emissions: air pollutants entering the atmosphere from other than a stack chimney, vent, or other functionally equivalent opening. Examples include vapors, dust, and fumes (40 CFR 51.301j).

Full Charge: the amount of refrigerant required for normal operating characteristics and conditions of the appliance as determined by using one of the following four methods or a combination of one of the following four methods (40 CFR 82.152): (1) the equipment manufacturers' determination of the correct full charge for the equipment; (2) determining the full charge by appropriate calculations based on component sizes, density of refrigerant, volume of piping, and all other relevant considerations; (3) the use of actual measurements of the amount of refrigerant added or evacuated from the appliance; or (4) the use of an established range based on the best available data, regarding the normal operating characteristics and conditions for the appliance, where the midpoint of the range will serve as the full charge and where records are maintained in accordance with 82.166(q).

Gasoline Carrier: any distributor who transports, stores, or is responsible for the transportation or storage of gasoline or diesel fuel without taking title to or otherwise having any ownership of the gasoline and without altering either the quality or quantity of the gasoline or diesel fuel (40 CFR 80.2).

Gasoline Distributor: any person who transports, stores, or is responsible for the transportation or storage of gasoline or diesel fuel at any point between any gasoline refinery or importer's facility and any retail outlet or wholesale purchaser consumer facility (40 CFR 80.2).

Halogenated Hazardous Air Pollutants Solvent (HAP): methylene chloride (CAS No. 75-09-2), perchloroethylene (CAS No. 127-18-4), trichloroethylene (CAS No. 79-01-6), 1,1,1-trichloroethane (CAS No. 71-55-6), carbon tetrachloride (CAS No. 56-23-5), and chloroform (CAS 67-66-3) (40 CFR 63.461).

Heat Input: heat derived from the combustion of fuel in a steam-generating unit that does not include the heat derived from preheated combustion air, recirculated flue gases, or exhaust gases from other sources (40 CFR 60.41c).

High-Pressure Appliance: an appliance that uses a refrigerant with a boiling point between −50 and 10°C (−58 and 50°F) at 29.9 in.Hg (76 cm) atmospheric pressure. This definition includes, but is not limited to, appliances using refrigerants -12, -22, -114, -500, or -502 (40 CFR 82.152).

Hourly Average: the arithmetic mean of all measurements taken during 1 h. At least two measurements must be taken during the hour [40 CFR 503.41(f)].

Household Waste: includes material discarded by single and multiple residential dwellings, hotels, motels, and other similar permanent or temporary housing (40 CFR 60.51a).

Idling Mode: the time period when a solvent-cleaning machine is not actively cleaning parts, and the sump heating coils are turned on (40 CFR 63.461).

Idling Mode Cover: any cover or solvent-cleaning machine design that shields the cleaning machine openings during the idling mode. A cover that meets this definition can also be used as a working mode cover if that definition is also met (40 CFR 63.461).

Immersion Cold-Cleaning Machine: a cold-cleaning machine in which the parts to be cleaned are immersed in the solvent. A remote reservoir cold-cleaning machine that is also an immersion cold-cleaning machine is considered an immersion cold-cleaning machine for the purposes of this subpart (40 CFR 63.461).

Incineration: in relation to sewage sludge, the combustion of organic matter and inorganic matter in sewage sludge by high temperatures in an enclosed device [40 CFR 503.41(g)].

Incinerator: any furnace used to burn solid waste for the purpose of reducing the volume of the waste by removing combustible matter (40 CFR 60.51).

Industrial Process Refrigeration: complex customized appliances used in the chemical, pharmaceutical, petrochemical, and manufacturing industries. This sector also includes industrial ice machines, appliances used directly in the generation of electricity, and ice rinks (40 CFR 82.152).

Industrial Process Shutdown: an industrial process or facility that temporarily ceases to operate or manufacture whatever is being produced at that facility (40 CFR 82.152).

Initial Verification Test: leak tests that are conducted as soon as practicable after the repair is completed (40 CFR 82.152).

In-Line-Cleaning Machine or Continuous-Cleaning Machine: a solvent-cleaning machine that uses an automated parts handling system, typically a conveyor, to automatically provide a continuous supply of parts for cleaning. These units are fully enclosed except for the conveyor inlet and exit portals. In-line-cleaning machines can be either cold- or vapor-cleaning machines (40 CFR 63.461).

Institutional Waste: materials discarded by hospitals, schools, nonmanufacturing activities at prisons, and government facilities (40 CFR 60.51a).

Large Municipal Waste Combustor Plant: a municipal waste combustor plant with a municipal waste combustor aggregate plant capacity for affected facilities that is >225 metric tons/day (250 t/day) of municipal solid waste (40 CFR 60.51a and 60.51b).

Lignite: coal that is classified as lignite A or B according to the ASTM standards (40 CFR 60.41a).

Lip Exhaust: a device installed at the top of the opening of a solvent-cleaning machine that draws in air and solvent vapor from the freeboard area and ducts the air and vapor away from the solvent-cleaning area (40 CFR 63.461).

Low-Loss Fitting: any device intended to establish a connection between hoses, appliances, or recovery or recycling machines that is designed to close automatically or to be closed manually when disconnected, minimizing the release of refrigerant from hoses, appliances, and recovery or recycling machines (40 CFR 82.152).

Low-Pressure Appliance: an appliance that uses a refrigerant with a boiling point above 10°C (50°F) at atmospheric pressure (29.9 in.Hg). This definition includes, but is not limited to, equipment utilizing refrigerants -11, -113, and -123 (40 CFR 82.152).

Major Maintenance, Service, or Repair: any maintenance, service, or repair involving the removal of any or all of the following appliance components: compressor, condenser, evaporator, or auxiliary heat exchanger coil (40 CFR 82.152).

Major Source: in relation to dry-cleaning facilities, any facility that emits or has the potential to emit more than 9.1 metric tons/year (10 t/year) of perchloroethylene to the atmosphere. In lieu of measuring a facility's potential to emit perchloroethylene emissions or determining a facility's potential to emit perchloroethylene emissions, a dry-cleaning facility is a major source if: it includes only dry-to-dry machines and has a total yearly perchloroethylene consumption >8000 l (2100 gal) or it includes only transfer machine systems or both dry-to-dry and transfer machine systems and has a total yearly perchloroethylene consumption >6800 l (1800 gal) (40 CFR 63.321).

Management Practice (MP): practices that, although not mandated by law, are encouraged to promote safe operating procedures.

Mass Burn Refractory Municipal Waste Combustor: a field-erected combustor that combusts municipal solid waste in a refractory wall furnace. Unless otherwise specified, this definition includes combustions with a cylindrical rotary refractory wall furnace (40 CFR 60.51a and 60.51b).

Mass Burn Rotary Waterwall Municipal Waste Combustor: a field-erected combustor that combusts municipal solid waste in a cylindrical rotary waterwall furnace (40 CFR 60.51a and 60.51b).

Materials Separation Plan: a plan that identifies both a goal and an approach to separate certain components of municipal solid waste for a given service area in order to make the separated materials available for recycling (40 CFR 60.51b).

Maximum Heat Input Capacity of a Steam-Generating Unit: determined by operating the facility at maximum capacity for 24 h and using the heat loss method described in Sections 5 and 7.3 of the ASME Power Test Codes 4.1 [see 40 CFR 60.17(h)] no later than 180 days after initial start-up of the facility and within 60 days after reaching the maximum production rate at which the facility will be operated (40 CFR 60.51a).

Modification: according to new source performance standards (NSPSs), any physical or operational change to an existing facility that results in an increase in the emission rate to the atmosphere of any pollutant to which a standard applies except: (1) maintenance, repair, and replacement that the administrator determines to be routine for a source category; (2) an increased production rate of an existing facility, if that increase can be accomplished without a capital expenditure on the facility; (3) an increase in the hours of operation; (4) use of an alternate fuel or raw material if, prior to the date any standard under this part becomes applicable to that source type, the existing facility was designed to accommodate that alternate use (A facility will be designed to accommodate an alternative fuel or raw material if that use could be accomplished under the facility's construction specifications as assessed prior to the change.); (5) the addition or use of any system or device whose primary function is the reduction of air pollutants, except when an emission control system is removed or replaced by a system that the administrator determines to be less than environmentally beneficial; and (6) the relocation or change in ownership of an existing facility (40 CFR 60.14).

Motor Vehicle Air-Conditioner (MVAC): any appliance that is an MVAC as defined in 40 CFR 82, subpart B (40 CFR 82.152).

Muck Cooker: a device for heating perchloroethylene-laden waste material to volatilize and recover perchloroethylene (40 CFR 63.321).

Municipal Solid Waste or Municipal-Type Solid Waste: household, commercial/retail, and/or institutional waste. Household waste includes materials discarded by single and multiple residential dwellings, hotels, motels, and other similar permanent or temporary housing establishments or facilities. Commercial/retail waste includes materials discarded by stores, offices, restaurants, warehouses, nonmanufacturing activities at industrial facilities, and other similar establishments or facilities. Institutional waste includes material discarded by schools, nonmedical waste discarded by hospitals, material discarded by nonmanufacturing activities at prisons and government facilities, and material discarded by other similar establishments. Household, commercial/retail, and institutional waste does not include used oil; sewage sludge; wood pallets; construction, renovation, and demolition wastes (which includes, but is not limited to, railroad ties and telephone poles); clean wood; industrial process or manufacturing wastes; medical waste; or motor vehicles. Household, commercial/retail, and institutional waste does include yard waste, refuse-derived fuel, and motor vehicle maintenance materials limited to vehicle batteries and tires except as specified in 40 CFR 60.50a(c) and 60.50b(g) (40 CFR 60.51a and 60.51b).

Municipal Waste Combustor Plant: one or more municipal waste combustor units at the same location for which construction, modification, or reconstruction commenced after December 20, 1989 and on or before September 20, 1994 (40 CFR 60.51a). One or more municipal waste combustor units at the same location for which construction, modification, or reconstruction commenced after September 20, 1994 (40 CFR 60.51b).

Municipal Waste Combustor Unit: any setting or equipment that combusts or gasifies municipal solid waste including, but not limited to, field-erected incinerators (with or without heat recovery), modular incinerators (starved air or excess air), boilers (i.e., steam-generating units), furnaces (whether suspension fired, grate-fired, mass-fired, air curtain incinerators, or fluidized bed-fired), and pyrolysis/combustion units. Municipal waste combustors do not include pyrolysis/combustion units located at plastics/rubber recycling units, internal combustion engines, gas turbines, or other combustion devices that combust landfill gases collected by landfill gas collection systems (40 CFR 60.51a and 60.51b).

MVAC-Like Appliance: mechanical vapor compression or open-drive compressor appliances used to cool the driver's or passenger's compartment of a nonroad motor vehicle. This definition includes the air-conditioning equipment found on agricultural or construction vehicles, but is not intended to cover appliances using HCFC-22 refrigerant (40 CFR 82.152).

New: in relation to a perchloroethylene dry-cleaning facility, a facility in which construction or reconstruction commenced on or after December 9, 1991 (40 CFR 63.321). In relation to solvent-cleaning machines, any machine whose construction or reconstruction commenced after November 29, 1993 (40 CFR 63.461).

Nitric Acid Production Unit: any facility producing nitric acid, which is 30 to 70% in strength, by either the pressure or atmospheric pressure process (40 CFR 60.70).

Normal Operating Characteristics or Conditions: temperature, pressure, fluid flow, speed, and other characteristics that would normally be expected for a given process load and ambient conditions during operation. Normal operating characteristics and conditions are marked by the absence of atypical conditions affecting the operation of the refrigeration appliance (40 CFR 82.152).

Normally Containing a Quantity of Refrigerant: containing the quantity of refrigerant within the appliance or appliance component when the appliance is operating with a full charge of refrigerant (40 CFR 82.152).

Opacity: the degree to which emissions reduce the transmission of light and obscure view of an object in the background (40 CFR 60.2).

Open Top, Vapor-Cleaning Machine: a batch solvent-cleaning machine that has its upper surface open to the air and boils solvents to create solvent vapor used to clean and/or dry parts (40 CFR 63.461).

Opening an Appliance: any service, maintenance, or repair on an appliance that would release class I or class II refrigerant from the appliance to the atmosphere unless the refrigerant were previously recovered from the appliance (40 CFR 82.152).

Particulate Matter Emissions: any airborne finely divided solid or liquid material, except uncombined water, emitted to the ambient air (40 CFR 60.2).

Perceptible Leaks: any perchloroethylene vapor or liquid leaks that are obvious from the odor of perchloroethylene; visual observation, such as pools or droplets of liquid; or the detection of gas flow by passing the fingers over the surface of the equipment (40 CFR 63.321).

Perchloroethylene Consumption: the total volume of perchloroethylene purchased based on purchase receipts or other reliable measures (40 CFR 63.321).

Petroleum Dry Cleaner: a dry-cleaning facility that uses petroleum solvent in a combination of washers, dryers, filters, stills, and settling tanks (40 CFR 60.621).

PM_{10}: particulate matter with an aerodynamic diameter less than or equal to a nominal 10 μm (40 CFR 58.1).

Process Stub: a length of tubing that provides access to the refrigerant inside a small appliance or room air-conditioner and can be resealed at the conclusion of repair or service (40 CFR 82.152).

Publication Rotogravure Printing: any number of rotogravure printing units capable of printing simultaneously on the same continuous web or substrate, and including any associated device for continuous cutting and folding the printed web, where the following sellable paper products are printed: catalogs; direct mail advertisements; display advertisements; magazines; miscellaneous advertisements including brochures, pamphlets, catalog sheets, circular folders, and announcements; newspapers; periodicals; and telephone and other directories (40 CFR 60.431).

Pyrolysis/Combustion Unit: a unit that produces gases, liquids, or solids through the heating of municipal solid waste. The gases, liquids, or solids produced are combusted, and the emissions are vented to the atmosphere (40 CFR 60.51b).

Reclaim Refrigerant: to reprocess refrigerant to at least the purity specified in the Air-Conditioning and Refrigeration Institute (ARI) Standard 700-1988, Specifications for Fluorocarbon Refrigerants (Appendix A to 40 CFR 82, subpart F), and to verify this purity using the analytical methodology prescribed in ARI Standard 700-1988. In general reclamation involves the use of processes or procedures available only at a reprocessing or manufacturing facility (40 CFR 82.152).

Reclaimer: a machine used to remove perchloroethylene from articles by tumbling them in a heated airstream (40 CFR 63.321).

Reconstruction: in relation to perchloroethylene dry-cleaners, replacement of a washer, dryer, or reclaimer, or replacement of any components of a dry-cleaning system, to such an extent that the fixed capital cost of the new components exceeds 50% of the fixed capital cost that would be required to construct a comparable new source (40 CFR 63.321).

Recover Refrigerant: to remove refrigerant in any condition from an appliance and to store it in an external container without necessarily testing or processing it in any way (40 CFR 82.152).

Recovery Efficiency: the percentage of refrigerant in an appliance that is recovered by a piece of recycling or recovery equipment (40 CFR 82.152).

Recycle Refrigerant: to extract refrigerant from an appliance and clean refrigerant for reuse without meeting all of the requirements for reclamation. In general recycled refrigerant is refrigerant that is cleaned using oil separation and single or multiple passes through devices, such as replaceable core filter-dryers, that reduce moisture, acidity, and particulate matter. These procedures usually are implemented at the field job site (40 CFR 82.152).

Refrigerated Condenser: a vapor recovery system into which an air–perchloroethylene gas–vapor stream is routed; the perchloroethylene is condensed by cooling the gas–vapor stream (40 CFR 63.321).

Refuse-Derived Fuel: the combustible or organic portion of municipal waste that has been separated out and processed for use as fuel (40 CFR 60.51a). A type of municipal solid waste produced by processing municipal solid waste through shredding and size classification, including all classes of refuse-derived fuel ranging from low-density fluff through densified and pelletized refuse-derived fuel (40 CFR 60.51b).

Remote Reservoir Cold-Cleaning Machine: any device in which liquid solvent is pumped to a sink-like work area that drains solvent back into an enclosed container while parts are being cleaned, allowing no solvent to pool in the work area (40 CFR 63.461).

Risk Specific Concentration: the allowable increase in the average daily ground level ambient air concentrations for a pollutant from the incineration of sewage sludge at or beyond the property line of the site where the sewage sludge incinerator is located [40 CFR 503.41(i)].

Room Enclosure: a stationary structure that encloses a transfer machine system and is vented to a carbon adsorber or an equivalent control device during operation of the transfer machine system (40 CFR 63.321).

Same Location: the same or contiguous property that is under common ownership or control, including properties that are separated only by a street, road, highway, or other public right of way. Common ownership or control includes properties that are owned, leased, or operated by the same entity, parent entity, subsidiary, subdivision, or any combination thereof, including any municipality or other governmental unit (40 CFR 60.51b).

Self-Contained Recovery Equipment: refrigerant recovery or recycling equipment that is capable of removing the refrigerant from an appliance without the assistance of components contained in the appliance (40 CFR 82.152).

Sewage Sludge Feed Rate: either the average daily amount of sewage sludge fired in all sewage sludge incinerators within the property line of the site, where the sewage sludge incinerators are located for the number of days in a 365-day period that each sewage sludge incinerator operates, or the average daily design capacity for all sewage sludge incinerators within the property line of the site where the sewage sludge incinerators are located [40 CFR 503.41(j)].

Sewage Sludge Incinerator: an enclosed device in which only sewage sludge and auxiliary fuel are fired [40 CFR 503.41(k)].

Small Appliance: any of the following products that are fully manufactured, charged, and hermetically sealed in a factory with 5 lb or less of refrigerant: refrigerators designed for home use, freezers designed for home use, room air-conditioners (including window air-conditioners and packaged terminal air-conditioners), packaged terminal heat pumps, dehumidifiers, under-the-counter ice makers, vending machines, and drinking water coolers (40 CFR 82.152).

Small Municipal Waste Combustor Plant: a municipal waste combustor plant with a capacity for affected facilities that is >35 metric tons/day, but ≤225 metric tons/day of municipal solid waste (40 CFR 60.51b).

Solvent/Air Interface: the location of contact between the concentrated solvent vapor layer and the air in a vapor-cleaning machine. This location of contact is defined as the in-line height of the primary condenser coils. For a cold-cleaning machine the solvent/air interface is the location of contact between the liquid solvent and the air (40 CFR 63.461).

Solvent-Cleaning Machine: any device or piece of equipment that uses halogenated HAP solvent liquid or vapor to remove soils from the surface of materials. The types of solvent-cleaning machines include, but are not limited to, batch vapor, in-line vapor, in-line cold, and batch cold solvent. Buckets, pails, and beakers with capacities of 7.6 l (2 gal) or less are not considered solvent-cleaning machines (40 CFR 63.461).

Stationary Gas Turbines: any simple cycle gas turbine, regenerative cycle gas turbine, or any gas turbine portion of a combined cycle steam/electric-generating system that *is not self-propelled. It may be* mounted on a vehicle for portability (40 CFR 60.331).

Steam-Generating Unit: any furnace, boiler, or other device used for combusting fuel for the purpose of producing steam (including fossil fuel-fired steam generators associated with combined cycle gas turbines); nuclear steam generators are not included (40 CFR 60.41a).

Suitable Replacement Refrigerant: for the purposes of 82.156(i)(7)(i), a refrigerant that is acceptable under section 612(c) of CAAA90 and all regulations promulgated under that section, compatible with other materials with which it may come into contact, and able to achieve the temperatures required for the affected industrial process in a technically feasible manner (40 CFR 82.152).

Sulfuric Acid Production Unit: any facility producing sulfuric acids by the contact process for burning elemental sulfur, alkylation acid, hydrogen sulfide, organic sulfides and mercaptans, or acid sludge, not including facilities where conversion to sulfuric acid is used primarily as a means of preventing emissions to the atmosphere of SO_2 or other sulfur compounds (40 CFR 60.81).

Superheated Vapor System: a system that heats the solvent vapor either passively or actively to a temperature above the solvent's boiling point. Parts are held in the superheated vapor before exiting the machine to evaporate the liquid solvent on them. Hot vapor recycle is an example of a superheated vapor system (40 CFR 63.461).

System-Dependent Recovery Equipment: refrigerant recovery equipment that requires the assistance of components contained in an appliance to remove the refrigerant from the appliance (40 CFR 82.152).

System Mothballing: the intentional shutting down of a refrigeration appliance undertaken for an extended period of time by the owners or operators of the facility, where the refrigerant has been evacuated from the appliance or the affected isolated section of the appliance, at least to atmospheric pressure (40 CFR 82.152).

Technician: any person who performs maintenance, service, or repair that could reasonably be expected to release class I or class II refrigerants from appliances, except for MVACs, into the atmosphere. Any person who performs disposal of appliances except for small appliances, MVAC, and MVAC-like equipment that could be reasonably expected to release class I or class II refrigerants from the appliances into the atmosphere. Technicians include, but are not limited to, installers, contractor employees, in-house service personnel, and owners (40 CFR 82.152).

Transfer Machine System: a multiple machine dry-cleaning operation in which washing and drying are performed in different machines. Examples include, but are not limited to, a washer and dryer, a washer and reclaimer, and a dry-to-dry machine and reclaimer (40 CFR 63.321).

Very High-Pressure Appliance: an appliance that uses a refrigerant with a boiling point below $-50°C$ ($-58°F$) at atmospheric pressure (29.9 in.Hg). This definition includes, but is not limited to, equipment utilizing refrigerants -13 and -503 (40 CFR 82.152).

Vapor-Cleaning Machine: a batch or in-line solvent-cleaning machine that boils liquid solvent, generating solvent vapor that is used as a part of the cleaning or drying cycle (40 CFR 63.461).

Very Low Sulfur Oil: an oil that contains no more than 0.5 weight percent sulfur or that, when combusted without SO_2 emission control, has an SO_2 emission rate ≤215 ng/J (0.5 lb/MBtu) heat input (40 CFR 60.41b).

VHAP Service: a piece of equipment that either contains or contacts a fluid (liquid or gas) that is at least 10% by weight a VHAP (40 CFR 61.241).

VOC Service: in relationship to fugitive emissions, VOC service begins when a piece of equipment contains or contacts a process fluid that is at least 10% VOC by weight (40 CFR 61.241).

Volatile Hazardous Air Pollutant (VHAP): a substance regulated under 40 CFR 61, subpart V, for which a standard for equipment leaks of the substance has been proposed and promulgated. Benzene and vinyl chloride are VHAPs (40 CFR 61.241).

Volatile Organic Compound (VOC): any compound of carbon, excluding CO, CO_2, carbonic acid, metallic carbides, carbonates, and ammonium carbonate, that participates in atmospheric photochemical reactions (40 CFR 51.100).

Voluntary Certification Program: a technician-testing program operated by a person before that person obtained approval of a technician certification program (40 CFR 82.152).

Washer: a machine used to clean articles by immersing them in perchloroethylene, including a dry-to-dry machine used with a reclaimer (40 CFR 63.321).

Water Separator: any device used to recover perchloroethylene from a water-perchloroethylene mixture (40 CFR 63.321).

Waterfall Furnace: a combustion unit having energy (heat) recovery in the furnace (i.e., radiant heat transfer section of the combustor) (40 CFR 60.51b).

Wholesale Purchaser-Consumer: any organization that is an ultimate consumer of gasoline or diesel fuel that purchases or obtains gasoline or diesel fuel from a supplier for use in motor vehicles and receives delivery of that product into a storage tank of at least 550 gal (2082 l) capacity substantially under the control of that organization (40 CFR 80.2).

Working Mode: the time period when the solvent-cleaning machine is actively cleaning parts (40 CFR 63.461).

Working Mode Cover: any cover or solvent-cleaning machine design that shields the cleaning machine openings from outside air disturbances while parts are being cleaned in the cleaning machine. A cover used during the working mode is opened only during parts entry and removal. A cover that meets this definition can also be used as an idling mode cover if that definition is also met (40 CFR 63.461).

Yard Waste: grass, grass clippings, bushes, shrubs, and clippings from bushes and shrubs that are generated by residential, commercial/retail, institutional, and/or industrial sources as a part of maintenance activities associated with yards or other private or public lands. Yard waste does not include construction, renovation, and demolition wastes that are exempt from the definition of municipal solid waste. Yard waste does not include clean wood (40 CFR 60.51b).

12.4 COMMUNITY RELATIONS

If a facility is of interest to the local community, and especially if an incident/accident occurs, information about the air dispersion situation and risk management plan will no doubt receive media coverage. Organizations will want to make sure that they have briefed local officials and other key stakeholders, such as neighbors, before this information appears in the news or before the stakeholders are contacted by media people seeking their reactions.

Organization managers can ensure that they are providing complete information to community leaders by briefing these people themselves. Community leaders may include local officials and other key stakeholders, such as neighbors. Other members of the community are likely to call community leaders for help in understanding the site's activities. Organizations should give community leaders the information necessary to answer general questions and to encourage residents to call Risk Management Planning (RMP) reporting organizations directly for more details.

A dialogue with the community will be most effective if it is started before an accident/incident occurs. The activities conducted will depend on the level of community interest, the visibility of facility operations, and other factors (such as the safety record of the facility, community perceptions about the type of chemicals being handled, and the number of people potentially affected in a worst-case scenario.

12.4.1 Notification

Public notification procedures must be developed with local emergency response groups that accommodate community concerns. When these procedures are developed, documentation will help reduce or eliminate the sense of dread that may develop for those people who may be potentially affected by an accident at the facility. Any special concerns given potential airborne or other environmental media contaminant release should be added to these procedures. Separate notification for schools and senior citizen facilities may ease community concerns for family and friends.

12.4.2 Fact Sheets

The preparation of consistently defined fact sheets of responses to the most likely asked questions will help to eliminate confusion during community outreach. Misperceptions about facility operations may grow when information is lacking, unavailable, or difficult for the community to interpret. Accidents, changes in the physical appearance of the facility, and seemingly unrelated events such as a dispute with organized labor often result in misinformation that spreads through the community that can either start or add to misperceptions. A short, preferably less than two pages, fact sheet that can be understood by the community serves two functions: it fills an information gap and provides something in writing.

12.4.3 Explaining Air Monitoring to the General Public

Communication to the public should include factual information and explanations. Records to review include the following:

- State and local air pollution control regulations
- Emissions inventory
- All air pollution source permits
- Plans and procedures applicable to air pollution control
- Emission monitoring records
- Opacity records

- Notices of violation (NOVs) from regulatory authorities
- Instrument calibration and maintenance records
- Reports/complaints concerning air quality
- Air emergency episode plan
- State and/or federal regulatory inspections
- Regulatory inspection reports
- Documentation of preventive measures or actions
- Results of air sampling at the conclusion of the response action
- Pollution prevention management plan
- Ozone-depleting chemical (ODC) inventory
- Training and certification records for employees who reclaim/recycle refrigerant

To the extent that off-site emissions are a concurrent issue, discussions should also include

- All air pollution sources (fuel burners, incinerators, VOC sources, etc.)
- Air pollution monitoring and control devices
- Air emission stacks
- Air intake vents

12.4.4 Employee Education

Inform employees of the purpose, content, and answers to these most likely to be asked questions about a facility or site. One of the primary sources of information about an organization to the community is its employees. If they are properly informed, the chances of misinformation and rumors will decrease.

Select a spokesperson who understands emergency response procedures and who is available locally to handle questions from citizen groups and the news media. This spokesperson should also be the initial point of contact for emergency providers whenever questions arise.

12.4.5 Public Accessibility

For some organizations there will be a high level of public interest in their organization and risk management plans. The level of public interest can depend on several factors, including your organization's safety record, the types and quantities of materials used, and prior level of interest in the organization by the community. Organizations that have a high level of public interest should consider means to increase public accessibility to information about ongoing risk management efforts. Dual training sessions with emergency providers and discussions at that time as to air-monitoring protocols are recommended.

12.4.6 Repository

Create a file containing the risk management plan, environmental permits, emergency response information, and general information about the organization. Select a location, preferably off-site, such as a library, where some information about the organization is available to the public.

12.4.7 Dialogue

For some organizations, setting up a citizens advisory panel (CAP) will be the best way to identify the concerns of the community. CAPs are most often used in larger communities where there are many different stakeholders and competing interests. CAPs can help organizations by providing a forum for gathering public opinion, providing accurate information, and resolving differences. The panels are usually represented by individuals from many different segments of the community.

Who should be part of this dialogue? The individuals and groups that should be initially contacted include the following:

- Adjacent property owners
- Administrators of organizations within the worst-case scenario distance: schools, nursing and senior citizen facilities, hospitals, day-care centers, and places of worship
- County board members
- Mayor and council members
- Public health agencies
- Civic and environmental groups
- Media

CAPs are one of many management initiatives that can be used to establish and maintain formal dialogue when

- A large number of residents are potentially affected by the operation
- Multiple communities are involved
- High interest in the facility is revealed through direct inquiries or news media coverage
- Misperceptions exist about risk posed to the community, site safety, operations, or other key issues

Glossary of Terms

Acclimatization. The process of becoming accustomed to new environmental conditions.

ACGIH. The American Conference of Governmental Industrial Hygienists is a professional organization devoted to worker health protection. The organization publishes *Threshold Limit Values for Chemical Substances in the Work Environment* and the *Documentation of TLVs*. The TLV booklet is one source that may be used in hazard determination.

Action Level. Term used by OSHA and NIOSH to express the level of toxicant that requires medical surveillance, usually one half the PEL.

Acute. An adverse effect on the human body with symptoms of high severity coming quickly to a crisis. Acute effects are normally the result of short-term exposures and short duration.

Administrative Controls. Methods of controlling employee exposures by job rotation, work assignment, or time periods away from the hazard.

AEC. Atomic Energy Commission. Now called the Nuclear Regulatory Commission in the U.S. Department of Energy.

Aerosols. Liquid droplets or solid particles dispersed in air with a fine enough particle size (0.01 to 100 μm) to remain so dispersed for a period of time.

AIHA. American Industrial Hygiene Association.

Air Monitoring. Sampling for and measuring pollutants in the atmosphere.

Air-Regulating Valve. An adjustable valve used to regulate airflow to the facepiece, helmet, or hood of a respirator.

Air-Purifying Respirator. Respirator that uses filters or sorbents to remove harmful substances from the air.

Air-Supplied Respirator. Respirator that provides a supply of breathable air from a clean source outside the contaminated work area.

Alpha-Emitter. A radioactive substance that gives off alpha particles.

Alpha Particle (alpha ray, alpha radiation). A small electrically charged particle of very high velocity thrown off by many radioactive materials, including uranium and radium. It is made up of two neutrons and two protons. Its electric charge is positive.

Annual Report on Carcinogens. A list of substances that are either known or anticipated as carcinogens. It is published by the National Toxicology Program (NTP).

ANSI. The American National Standards Institute is a voluntary membership organization (run with private funding) that develops consensus standards nationally for a wide variety of devices and procedures.

Approved. Tested and listed as satisfactory by an authority having jurisdiction, such as the U.S. Department of Health and Human Services, NIOSH-MSHA, or the U.S. Department of Agriculture.

Arc Welding.   One form of electrical welding using either uncoated or coated rods.

Arc-Welding Electrode.   A component of the welding circuit through which current is conducted between the electrode holder and the arc.

Asbestos.   A hydrated magnesium silicate in fibrous form.

Asbestosis.   A disease of the lungs caused by the inhalation of fine airborne fibers of asbestos.

Asphyxia.   Suffocation from a lack of oxygen. Chemical asphyxia is produced by a substance, such as carbon monoxide, that combines with hemoglobin to reduce the blood's capacity to transport oxygen. Simple asphyxia is the result of exposure to a substance, such as methane, that displaces oxygen.

Attenuate.   To reduce in amount or intensity. Usually refers to noise or ionizing radiation.

Attenuation.   The reduction of the intensity at a designated first location as compared with intensity at a second location, which is farther from the source.

Audiogram.   A record of hearing loss or hearing level measured at several different frequencies—usually 500–6,000 Hz. The audiogram may be presented graphically or numerically. Hearing level is shown as a function of frequency.

Audiologist.   A person with graduate training in the specialized problems of hearing and deafness.

Audiometer.   A signal generator or instrument for measuring objectively the sensitivity of hearing in decibels referred to as audiometric zero. Pure tone audiometers are standard instruments for industrial use for audiometric testing.

Audiometric Technician.   A person who is trained and qualified to administer audiometric examinations.

Autoignition Temperature.   The lowest temperature at which a flammable gas or vapor-air mixture will ignite from its own heat source or a contacted heated surface without a spark or flame. Vapors and gases will spontaneously ignite at a lower temperature in oxygen than in air, and their autoignition temperature may be influenced by the presence of catalytic substances.

Beta particle (beta radiation).   A small electrically charged particle thrown off by many radioactive materials. It is identical with the electron. Beta particles emerge from radioactive material at high speeds.

Boiling Point.   The temperature at which the vapor pressure of a liquid equals atmospheric pressure.

Breathing Zone.   Imaginary globe of 2-ft radius surrounding the head.

Cancer.   A cellular tumor whose natural course is fatal that is usually associated with the formation of secondary tumors.

Capture Velocity.   Air velocity at any point in front of the exhaust hood necessary to overcome opposing air currents and to capture the contaminated air by forcing it to flow into the exhaust hood.

Carcinogen.   A substance that causes cancer. Cancer is characterized by the proliferation of abnormal cells, sometimes in the form of a tumor.

CAS Number.   An identification number assigned by the Chemical Abstracts Service (CAS) of the American Chemical Society. The CAS Number is used in various databases, including Chemical Abstracts, for identification and information retrieval.

Caustic.   Something that strongly irritates, burns, corrodes, or destroys living tissue.

Ceiling Limit (C).   An airborne concentration of a toxic substance in the work environment that should never be exceeded.

CFR.   Code of Federal Regulations—the collection of rules and regulations originally published in the *Federal Register* by various governmental departments and agencies.

OSHA regulations are found in 29 CFR; EPA regulations are in 40 CFR; and Department of Transportation regulations are in 49 CFR.

Combustible Liquids.   Liquids having a flash point at or above 37.8°C (100°F).

Comfort Ventilation.   Airflow intended to maintain the comfort of room occupants (heat, humidity, and odor).

Compressed Gas (OSHA).   A gas or mixture of gases having, in a container, an absolute pressure exceeding 40 psi at 70°F (21.1°C). A gas or mixture of gases having, in a container, an absolute pressure exceeding 104 psi at 130°F (54.4°C) regardless of the pressure at 70°F (21.1°C). A liquid having a vapor pressure exceeding 40 psi at 100°F (37.8°C) as determined by ASTM D-323-72.

Confined Space.   According to NIOSH's *Criteria for a Recommended Standard: Working in Confined Spaces*, a space that by design has limited openings for entry and exit, unfavorable natural ventilation that could contain or produce dangerous air contaminants, and not intended for continuous employee occupancy. Confined spaces include, but are not limited to, storage tanks, compartments of ships, process vessels, pits, silos, vats, degreasers, reaction vessels, boilers, ventilation and exhaust ducts, sewers, tunnels, underground utility vaults, and pipelines. A confined space entry would be considered a nonroutine task.

Corrective Lens.   A lens ground to the wearer's individual optical prescription.

Corrosion.   A physical change, usually deterioration or destruction, caused by chemical or electrochemical action as contrasted with erosion caused by mechanical action.

Corrosive.   A substance that causes visible destruction or permanent changes in human skin tissue at the site of contact.

Cryogenics.   The field of science dealing with the behavior of matter at very low temperatures.

Daughter.   As used in radioactivity, the product nucleus or atom resulting from the decay of the precursor or parent.

dBA.   Sound level in decibels read on the A-scale of a sound-level meter. The A-scale discriminates against very low frequencies (as does the human ear) and is, therefore, better for measuring general sound levels.

Decibel (dB).   A unit used to express sound power level (L_w). Sound power is the total acoustic output of a sound source in watts. Sound power level, in decibels, is $L_w = 10$ log $W - W_o$, where W is the sound power of the source and W_o is the reference sound power.

Decontaminate.   To make safe by eliminating poisonous or otherwise harmful substances, such as noxious chemicals or radioactive material.

Dermatitis.   Inflammation of the skin from any cause.

Differential Pressure.   The difference in static pressure between two locations.

Dilution.   The process of increasing the proportion of solvent or diluent (liquid) to solute or particulate matter (solid).

DOP.   Dioctyl phthalate, a powdered chemical that can be aerosolized to an extremely uniform size, i.e., 0.3 μm, for a major portion of any sample.

Dose.   A term used to express the amount of a chemical or ionizing radiation energy absorbed in a unit volume or an organ or individual. Dose rate is the dose delivered per unit of time (*see also* Roentgen, Rad, Rem). A term used to express the amount of exposure to a chemical substance.

Dosimeter (dose meter).   An instrument used to determine the full-shift exposure a person has received to a physical hazard.

DOT.   U.S. Department of Transportation.

Droplet. A liquid particle suspended in a gas. The liquid particle is generally of such size and density that it settles rapidly and only remains airborne for an appreciable length of time in a turbulent atmosphere.

Dry-Bulb Thermometer. An ordinary thermometer, especially one with an unmoistened bulb not dependent on atmospheric humidity. The reading is the dry-bulb temperature.

Duct. A conduit used for conveying air at low pressures.

Dusts. Solid particles generated by handling, crushing, grinding, rapid impact, detonation, and decrepitation of organic or inorganic materials, such as rock, ore, metal, coal, wood, and grain. Dusts do not tend to flocculate, except under electrostatic forces; they do not diffuse in air, but settle under the influence of gravity.

Engineering Controls. Methods of controlling employee exposures by modifying the source or reducing the quantity of contaminants released into the workroom environment.

EPA. U.S. Environmental Protection Agency.

Epidermis. The superficial scarfskin or upper (outer) strata of skin.

Exhalation valve. A device that allows exhaled air to leave a respirator and prevents outside air from entering through the valve.

Exhaust Ventilation. The removal of air usually by mechanical means from any space. The flow of air between two points is due to the occurrence of a pressure difference between the two points. This pressure difference will force air to flow from the high pressure zone to the low pressure zone.

Exposure. Contact with a chemical, biological, or physical hazard.

Eyepiece. Gas-tight, transparent window(s) in a full facepiece through which the wearer may see.

Facepiece. That portion of a respirator that covers the wearer's nose and mouth in a half-mask facepiece or nose, mouth, and eyes in a full facepiece. It is designed to make a gas-tight or dust-tight fit with the face and includes the headbands, exhalation valve(s), and connections for the air-purifying device or respirable-gas source or both.

Face Velocity. Average air velocity into the exhaust system measured at the opening into the hood or booth.

Federal Register. Publication of U.S. government documents officially promulgated under the law, documents whose validity depends on such publication. It is published on each day following a government working day. It is, in effect, the daily supplement to the *Code of Federal Regulations.*

Filter. (1) A device for separating components of a signal on the basis of its frequency. It allow components in one or more frequency bands to pass relatively unattenuated, while effectively attenuating components in other frequency bands. (2) A fibrous medium used in respirators to remove solid or liquid particles from the airstream entering the respirator. (3) A sheet of material that is interposed between a patient and the source of X-rays to absorb a selective part of the X-rays. (4) A fibrous or membrane medium used to collect dust, fume, or mist air samples.

Filter Efficiency. The efficiency of various filters can be established on the basis of entrapped particles, i.e., collection efficiency, or on the basis of particles passed through the filter, i.e., penetration efficiency.

Flammable Aerosol. An aerosol that is required to be labeled as "Flammable" under the Federal Hazardous Substances Labeling Act (15 USC 1261).

Flammable Limits. Flammables have a minimum concentration below which propagation of flame does not occur on contact with a source of ignition. This point is known as the LEL. There is also a maximum concentration of vapor or gas in air above which propagation of flame does not occur. This point is known as the UEL. These units are expressed in the percent of gas or vapor in air by volume.

Flammable Liquid. Any liquid having a flash point below 37.8°C (100°F).

Flammable Range. The difference between the LEL and the UEL, expressed in terms of the percentage of vapor or gas in air by volume, and often referred to as the "explosive range."

Frequency. Rate at which pressure oscillations are produced. A subjective characteristic of sound related to frequency is pitch.

Fume. Airborne particulate formed by the evaporation of solid materials, e.g., metal fumes emitted during welding. Usually less than 1 μm in diameter.

Gas Chromatography. A gaseous detection technique that involves the separation of mixtures by passing them through a column that will enable the components to be suspended for varying periods of time before they are detected or recorded.

Globe Thermometer. A thermometer set in the center of a metal sphere that has been painted black in order to measure radiant heat.

Grab Sample. A sample that is taken within a very short time period. The sample is taken to determine the constituents at a specific time.

Gravimetric Method. A procedure dependent on the formation or use of a precipitate or residue that is weighed to determine the concentration of a specific contaminant in a previously collected sample.

Hearing Conservation. The prevention or minimization of noise-induced deafness through the use of hearing protection devices; the control of noise through engineering methods, annual audiometric tests, and employee training.

Heat Cramps. Painful muscle spasms as a result of exposure to excess heat.

Heat Exhaustion. A condition usually caused by loss of body water because of exposure to excess heat. Symptoms include headache, tiredness, nausea, and sometimes fainting.

Heat Stress. Relative amount of thermal strain from the environment.

Heat Stroke. A serious disorder resulting from exposure to excess heat. It results from sweat suppression and increased storage of body heat. Symptoms include hot dry skin, high temperature, mental confusion, convulsions, and coma. It is fatal if not treated promptly.

Helmet. A device that shields the eyes, face, neck, and other parts of the head.

HEPA Filter. A disposable, extended medium, dry type filter with a particle removal efficiency of no less than 99.97% of 0.3 μm particles.

Hertz. The frequency measured in cycles per second (1 cps = 1 Hz).

Humidity. (1) Absolute humidity is the weight of water vapor per unit volume, pounds per cubic foot, or grams per cubic centimeter. (2) Relative humidity is the ratio of the actual partial vapor pressure of the water vapor in a space to the saturation pressure of pure water at the same temperature.

IARC. International Agency for Research on Cancer.

IDLH. Immediately dangerous to life or health.

Inert (Chemical). Not having active properties.

Inert Gas. A gas that does not normally combine chemically with the base metal or filler metal.

Infrared. Those wavelengths (10^{-4} to 10^{-1} cm) of the electromagnetic spectrum longer than those of visible light and shorter than radio waves.

Infrared Radiation. Electromagnetic energy with wavelengths from 770 nm to 12,000 nm.

Ionizing Radiation. Electrically charged or neutral particles or electromagnetic radiation that will interact with gases, liquids, or solids to produce ions. There are five major types: alpha, beta, X-ray, gamma, and neutrons.

LC$_{50}$. Lethal concentration that will kill 50% of test animals within a specified time. *See also* LD$_{50}$.

LD$_{50}$.   The dose required to produce death in 50% of the exposed species within a specified time.

Local Exhaust Ventilation.   A ventilation system that captures and removes the contaminants at the point they are being produced before they escape into the workroom air.

Localized.   Restricted to one spot or area in the body and not spread all through it (*see also* systemic).

Lower Explosive Limit (LEL).   The lower limit of flammability of a gas or vapor at ordinary ambient temperatures expressed in the percent of the gas or vapor in air by volume. This limit is assumed constant for temperatures up to 120°C (250°F). Above this temperature it should be decreased by a factor of 0.7 because explosibility increases with higher temperatures.

Makeup Air.   Clean, tempered outdoor air supplied to a work space to replace air removed by exhaust ventilation or some industrial process.

Manometer.   Instrument for measuring pressure; essentially a U-tube partially filled with a liquid (usually water, mercury, or a light oil) so constructed that the amount of displacement of the liquid indicates the pressure being exerted on the instrument.

Maximum Use Concentration (MUC).   The product of the protection factor of the respiratory protection equipment and the PEL.

Medical Surveillance.   Many OSHA-regulated chemicals have undesirable health-related effects; therefore OSHA requires that the employer conduct medical surveillance on employees to maintain chemical exposure within acceptable limits.

Mesothelioma.   Cancer of the membranes that line the chest and abdomen.

Micrometer (Micron).   A unit of length equal to 10^{-4} cm, approximately 1/25,000 of an inch.

Milligrams per Cubic Meter.   Unit used to measure air concentrations of dusts, gases, mists, and fumes.

Mists.   Suspended liquid droplets generated by condensation from the gaseous to the liquid state or by breaking up a liquid into a dispersed state, such as by splashing, foaming, or atomizing. Mist is formed when a finely divided liquid is suspended in air.

Mppcf.   Million particles per cubic foot.

MSHA.   The Mine Safety and Health Administration is a federal agency that regulates the safety and health of the mining industry.

MSDS.   Material safety data sheet.

NFPA.   The National Fire Protection Association is a voluntary membership organization whose aim is to promote and improve fire protection and prevention. The NFPA publishes 16 volumes of codes known as the National Fire Codes.

NIOSH.   The National Institute for Occupational Safety and Health is a federal agency that conducts research on health and safety concerns, tests and certifies respirators, and trains occupational health and safety professionals.

Noise-Induced Hearing Loss.   The terminology used to refer to the slowly progressive inner ear hearing loss that results from exposure to continuous noise over a long period of time as contrasted to acoustic trauma or physical injury to the ear.

NRC.   Nuclear Regulatory Commission of the U.S. Department of Energy.

NTP.   National Toxicology Program.

Nuisance Dust.   Dusts that have a long history of little adverse effect on the lungs and that do not produce significant organic disease or toxic effects when exposures are kept under reasonable control.

Odor.   That property of a substance that affects the sense of smell.

Odor Threshold.   The minimum concentration of a substance at which a majority of test subjects can detect and identify the characteristic odor of a substance.

Olfactory.   Associated with the sense of smell.

OSHA.   U.S. Occupational Safety and Health Administration.

Overexposure.   Exposure beyond the specified limits.

Oxidizing Agent.   A chemical that gives off free oxygen in a chemical reaction.

Particle.   A small discrete mass of solid or liquid matter.

Particulate.   A particle of solid or liquid matter.

Particulate Matter.   A suspension of fine solid or liquid particles in air, such as dust, fog, fume, mist, smoke, or sprays. Particulate matter suspended in air is commonly known as an aerosol.

Penetration.   The passage of a chemical through an opening in a protective material. Holes and rips in protective clothing can allow penetration as can stitch holes, space between zipper teeth, and open jacket and pant cuffs.

Permeation.   The passage of a chemical through a piece of clothing on a molecular level. If a piece of clothing is permeated, the chemical may collect on the inside, increasing the chance of skin contact with that chemical. Permeation is independent of degradation. Permeation may occur even though the clothing may show no signs of degradation.

Permissible Exposure Limit (PEL).   An exposure limit that is published and enforced by OSHA as a legal standard.

Personal Protective Equipment (PPE).   Devices worn by the worker to protect against hazards in the environment. Respirators, gloves, and hearing protectors are examples of PPE.

ppb.   Parts per billion.

ppm.   Parts of vapor, gas, or other contaminants per million parts of air by volume.

Psychrometer.   An instrument consisting of wet and dry bulb thermometers for measuring relative humidity.

Pulmonary.   Pertaining to the lungs.

RAD.   Radiation absorbed dose.

Radioactive.   The property of an isotope or element that is characterized by spontaneous decay to emit radiation.

Radioactivity.   Emission of energy in the form of alpha, beta, or gamma radiation from the nucleus of an atom. Always involves the change of one kind of atom into a different kind. A few elements, such as radium, are naturally radioactive. Other radioactive forms are induced.

Reagent.   Any substance used in a chemical reaction to produce, measure, examine, or detect another substance.

Respirator.   A device to protect the wearer from inhalation of harmful contaminants.

Respiratory System.   Consists of (in descending order) the nose, mouth, nasal passages, nasal pharynx, pharynx, larynx, trachea, bronchi, bronchioles, air sacs (alveoli) of the lungs, and muscles of respiration.

RTECS.   Registry of Toxic Effects of Chemical Substances. This NIOSH publication is one of the information sources OSHA recommends for hazard determination. RTECS provides data on toxicity for over 50,000 different chemicals. It has an extensive cross-reference listing trade names and synonyms. It is available as hard copy, computer tape, microfiche, and on-line through the National Library of Medicine.

Safety Can.   An approved container, of not more than 19 l (5 gal) capacity, having a spring-closing lid and spout cover, and designed so that it will safely relieve internal pressure when subjected to fire exposure.

Sampling.   A process consisting of the withdrawal or isolation of a fractional part of a whole. In air analysis the separation of a portion of an ambient atmosphere with subsequent analysis to determine concentration.

Sealed Source. A radioactive source sealed in a container or having a bonded cover, where the container or cover has sufficient mechanical strength to prevent contact with and dispersion of the radioactive material under the conditions of use and wear for which it is designed.

SCBA. Self-contained breathing apparatus.

Short-Term Exposure Limit (STEL). ACGIH-recommended exposure limit. Maximum concentration to which workers can be exposed for a short period of time (15 min) for only four times throughout the day with at least 1 h between exposures.

Sound Level. A weighted sound pressure level, obtained by the use of metering characteristics and the weighting A, B, or C specified in ANSI S1.4.

Sound Level Meter and Octave-Band Analyzer. Instruments for measuring sound pressure levels in decibels referenced to 0.0002 μbars. Readings can also be made in specific octave bands, usually beginning at 75 Hz and continuing through 10,000 Hz.

Sound Pressure Level (SPL). The level, in decibels, of a sound is 20 times the logarithm to the base 10 of the ratio of the pressure of this sound to the reference pressure. The reference pressure must be explicitly stated.

Specific Gravity. The ratio of the mass of a unit volume of a substance to the mass of the same volume of a standard substance at a standard temperature. Water at 4°C (39.2°F) is the standard usually referred to for liquids; dry air (at the same temperature and pressure as the gas) is often taken as the standard substance for gases.

Spectrography (Spectral Emission). An instrumental method for detecting trace contaminants utilizing the formation of a spectrum by exciting the contaminants under study by various means, causing characteristic radiation to be formed, which is dispersed by a grating or a prism and photographed.

Spectrophotometer. An instrument used for comparing the relative intensities of the corresponding colors produced by chemical reactions.

Stability. A measure of the ability of a substance to be handled and stored without undergoing unwanted chemical changes.

Standard Industrial Classification (SIC) Code. Classification system for places of employment according to the major type of activity.

Static Pressure. The potential pressure exerted in all directions by a fluid at rest. For a fluid in motion, it is measured in a direction normal (at right angles) to the direction of flow, thus it shows the tendency to burst or collapse the pipe. When added to velocity pressure, it gives total pressure.

Stressor. Any agent or thing causing a condition of stress.

Suspected Carcinogen. A material that is believed to be capable of causing cancer, but for which there is limited empirical evidence.

Symptom. Any bit of evidence from a patient indicating illness; the subjective feelings of the patient.

Synergistic. Pertaining to an action of two or more substances, organs, or organisms to achieve an effect that is greater than the additive effects of each alone.

Synonym. Another name by which the same chemical may be known.

Systemic. Spread throughout the body, affecting all body systems and organs, not localized in one spot or area.

Temperature. The condition of a body that determines the transfer of heat to or from other bodies. Specifically, it is a manifestation of the average translational kinetic energy of the molecules of a substance due to heat agitation.

Temperature, Dry-Bulb. The temperature of a gas or mixture of gases indicated by an accurate thermometer after correction for radiation.

Temperature, Wet-Bulb. The temperature at which liquid or solid water, by evaporating into air, can bring the air to saturation adiabatically at the same temperature. Wet-bulb temperature (without qualification) is the temperature indicated by a wet-bulb psychrometer.

Temporary Threshold Shift (TTS). The hearing loss suffered as the result of noise exposure, all or part of which is recovered during an arbitrary period of time when one is removed from the noise. It accounts for the necessity of checking hearing acuity at least 16 h after a noise exposure.

Teratogen. An agent or substance that may cause physical defects in the developing embryo or fetus when a pregnant female is exposed to that substance.

Threshold Limit Value (TLV). A TWA concentration under which most people can work consistently for 8 h a day, day after day, with no harmful effects. A table of these values and accompanying precautions is published annually by ACGIH.

Toxicity. A relative property of a chemical agent that refers to a harmful effect on some biologic mechanism and the condition under which this effect occurs.

Toxin. A poisonous substance that is derived from an organism.

TWA. Time-weighted average.

Ultraviolet. Those wavelengths (10^{-5} to 10^{-6} cm) of the electromagnetic spectrum that are shorter than those of visible light and longer than X-rays.

Unstable Radioactive Elements. The tendency of all radioactive elements is to emit particles and decay to form other elements.

Unstable (Reactive) Liquid. A liquid that in the pure state or as commercially produced or transported will vigorously polymerize, decompose, condense, or become self-reactive under conditions of shock, pressure, or temperature.

Upper Explosive Limit (UEL). The highest concentration (expressed in percent vapor or gas in the air by volume) of a substance that will burn or explode when an ignition source is present.

Vapor Pressure. Pressure (measured in pounds per square inch absolute [psia]) exerted by a vapor. If a vapor is kept in confinement over its liquid so that the vapor can accumulate above the liquid (the temperature being held constant), the vapor pressure approaches a fixed limit called the maximum (or saturated) vapor pressure, dependent only on the temperature and the liquid.

Vapor. The gaseous form of a substance that is normally in the solid or liquid state (at room temperature and pressure). A vapor can be changed back to the solid or liquid state either by increasing the pressure or decreasing the temperature. Vapors also diffuse. Evaporation is the process by which a liquid is changed into the vapor state and mixed with surrounding air. Solvents with low boiling points will volatilize readily.

Velocity, Face. The inward air velocity in the plane of opening into an enclosure.

Velometer. A device for measuring air velocity.

Ventilation. One of the principal methods to control health hazards—it causes fresh air to circulate to replace foul air simultaneously removed.

Ventilation, Dilution. Airflow designed to dilute contaminants to acceptable levels.

Ventilation, Mechanical. Air movement caused by a fan or other air moving device.

Ventilation, Natural. Air movement caused by wind, temperature difference, or other nonmechanical factors.

Volatility. The tendency or ability of a liquid to vaporize. Such liquids as alcohol and gasoline, because of their well-known tendency to evaporate rapidly, are called volatile liquids.

Volume Flow Rate. The quantity (measured in units of volume) of a fluid flowing per unit of time, as cubic feet per minute, gallons per hour, or cubic meters per second.

Welding Types.   The several types of welding are electric arc-welding, oxyacetylene welding, spot welding, and inert or shielded gas welding utilizing helium or argon.

Welding Rod.   A rod or heavy wire that is melted and fused into metals in arc welding.

Wet-Bulb Globe Temperature Index.   An index of the heat stress in humans when work is being performed in a hot environment.

Wet-Bulb Temperature.   Temperature as determined by the wet-bulb thermometer or a standard sling psychrometer or its equivalent. This temperature is influenced by the evaporation rate of water, which in turn depends on the humidity (amount of water vapor) in the air.

Wet-Bulb Thermometer.   A thermometer whose bulb is covered with a cloth saturated with water.

X-ray.   Highly penetrating radiation similar to gamma rays. Unlike gamma rays X-rays do not come from the nucleus of the atom, but from the surrounding electrons. They are produced by electron bombardment. When these rays pass through an object, they give a shadow picture of the denser portions.

X-ray Diffraction.   All crystals act as three-dimensional gratings for X-rays; the pattern of diffracted rays is characteristic for each crystalline material. This method is of particular value in determining the presence or absence of crystalline silica in an industrial dust.

Index

Milton Keynes UK
Ingram Content Group UK Ltd.
UKHW051948071024
449327UK00026B/2227